EFFECTIVE SCHOOL LIBRARIANSHIP

Successful Professional Practices from Librarians
Around the World

Funding for this book project was partially provided by the Faculty of Library, Information & Media Science, University of Tsukuba (Japan).

EFFECTIVE SCHOOL LIBRARIANSHIP

Volume II

Successful Professional Practices from Librarians
Around the World

Dr. Patrick Lo
Heather Rogers
Dr. Dickson K.W. Chiu

Apple Academic Press Inc.
3333 Mistwell Crescent
Oakville, ON L6L 0A2
Canada

Apple Academic Press Inc.
9 Spinnaker Way
Waretown, NJ 08758
USA

ISBN 13: 978-1-77463-529-2 (pbk)
ISBN 13: 978-1-77188-658-1 (hbk)
Successful Professional Practices from Librarians around the World (2-volume set)
ISBN 13: 978-1-77188-656-7 (hbk)

Library and Archives Canada Cataloguing in Publication

Lo, Patrick, author
Effective school librarianship : successful professional practices from librarians around the world / Dr. Patrick Lo, Heather Rogers, Dr. Dickson K.W. Chiu.

Includes bibliographical references and indexes.
Issued in print and electronic formats.
ISBN 978-1-77188-656-7 (set : hardcover).--ISBN 978-1-77188-657-4 (v. 1 : hardcover).--ISBN 978-1-77188-658-1 (v. 2 : hardcover).-- ISBN 978-1-315-14957-8 (PDF)

1. School librarians--Interviews. 2. School libraries--Case studies. 3. Library science--Case studies.
I. Chiu, Dickson K. W., 1966-, author
II. Rogers, Heather, 1990-, author III. Title.

Z675.S3L6 2018	027.8	C2018-900486-X	C2018-900405-3

CIP data on file with US Library of Congress

CONTENTS

FOREWORD

I feel honored to have been asked to write a foreword for this book. Its title, *Effective School Librarianship: Successful Professional Practices from Librarians around the World*, is intriguing, How do school librarians at public, private, and international schools of developing, emerging, and developed countries throughout the world manage to inspire children to read and improve the quality of their education? Are there some hidden secrets which could help us to make our work in school librarianship more effective?

This book, which is fascinating to read, talks about problems which occur and how these are being solved. The authors have collected interviews from school librarians throughout the world. Some of these people work under very difficult circumstances.

Interviewees have mentioned a multitude of "secrets" of their successful work, however the following themes occur constantly in the interviews.

- Many of the interviewees take pleasure in their work—they are people with passion. They speak with great concern and sympathy for the students who use the school library.
- There is a need for clear guidelines which should be applicable to school libraries throughout the world, at many different levels. This problem may have been solved to some extent by the publication of the *IFLA School Library Guidelines*, 2nd edition, which was published in June 2015 (http://www.ifla.org/files/assets/school-libraries-resource-centers/publications/ifla-school-library-guidelines.pdf). This document needs to be publicized.
- Some school librarians are highly trained and have not only a Library and Information Science (LIS) qualification but also a teaching qualification. Others have no official qualifications at all. Some are volunteers. Nevertheless, these people are successful because they provide a vital service to students and members of the entire school and local community.

- School librarians encourage students throughout the world to become literate, and encourage a reading culture in some countries where this does not exist. In some cases, the school libraries, in turn, affect the well-being of the child's family, giving them access to books and reading materials, which they did not previously have.
- School librarians co-operate with teachers to achieve the aims of the library. They also talk about the need for strong co-operation between school libraries and public libraries (as equal partners).
- School librarians often give students one-to-one attention, helping them to select a book which they will enjoy reading (either a novel or a book containing information for a project). Students are taught problem-solving skills.
- School librarians do their best to provide a physical and/or digital space where students can learn to read and broaden their horizons. Some of these locations are comfortable (state-of-the art) spaces with large physical and digital collections and technologies, while those at the other end of the scale may not even have an actual library room, let alone shelving, seating, and worktables. In others, these provisions are insufficient for the number of children attending the school. Some school libraries do not even have electricity, let alone a connection to the Internet. In these circumstances, children are provided with reading materials from classroom libraries or boxes containing books.
- The library provides a safe place where students, especially girls, can study and work on school assignments.
- Some interviewees have spoken about the need for multilingual, multicultural school libraries, where students (often refugees or migrants) make of just one library. Materials in different languages, providing different cultural perspectives, are provided. These aspects are very important in our dynamic, globalized world. The emergence of global literacy skills is also relevant in this context.
- Many librarians have mentioned the importance of the use of technology in their work, and in teaching students to use this technology in a correct way. School librarians create learning environments that are enhanced by technology.
- Some librarians commented on the need for training at many different levels, especially in the use of (new) technologies.

Furthermore, some schools and other organizations, in countries where there are very few books to read and hardly any libraries, still struggle with the establishment of a reading culture in their communities. In 2012, for example, Boelens, Boekhorst and Mangale completed a research study into 19 public primary schools in rural Kenya[1] and the attempts made by school leaders, teachers, and parents to establish a reading culture and school libraries in these schools. In the Australian and African Section of this book, Katherine Shaw describes the lack of a reading culture in D.R. Congo, where books are precious—a real luxury. Students in schools are often taught by rote, repeating texts dictated by teachers.

In conclusion, I would like to congratulate the authors for collecting such an interesting selection of interviews. It is my hope that after reading this book, educators, teachers, and librarians as well as members of the general public will have a better understanding of school librarianship across the world and will be inspired to co-operate with each other in many different ways, assisting those who desperately need help and support. This would be in the best interest of the children whom they serve and it relates to their duty of care as educators.

Dr. Helen Boelens
(International school library researcher and consultant,
Former Chair, IASL Research SIG,
The Netherlands)

[1] Boelens, H.; Boekhorst, A.; Mangale D. *School Libraries for 19 Public Schools in Rural Kenya: a Pilot Study*. 2012 111 p. http://www.albertkb.nl/mediapool/60/608240/data/Kenya-School-Libraries-Report-2012-2.pdf.

FOREWORD

Effective School Librarianship: Successful Professional Practices from Librarians around the World is an exceptional collection of scholarly reports and professional guidelines, timely delivered and significantly informing the readers about developments in school librarianship. It is my pleasure to write this Forward, and from a perspective of a teacher and an education technologists, give strongest endorsement to this line of intellectual debate in respect to the roles of modern libraries and teacher librarians in the 21st century schools.

Rapid developments in technologies and corresponding paradigm shifts that these create, present both, opportunities and challenges for education—not only for school libraries and teacher librarians, but all the way down to classrooms, teachers and students. As traditional school books are becoming enhanced and even replaced with e-books, digital learning resources, interactive and visual representations, augmented reality via technologies such as mobile devices, as access to information and people has been transformed with the Internet, social networking and data sources, and as technology tools, such as, those for media production, creative expressions and analytical processing are becoming easy to use and accessibly to human information activities, how and what we teach is changing (Churchill, 2017)[1]. Activities required for today's education need to be learning-centered, incorporate the above noted possibilities, and engage students in collaborative work in projects, inquiries, problem solving and research. To achieve these, learning requires not only access to information and technology, but also environment, 'coaches' for students to enhance their new literacies, and transformative leaders for teachers to support change in their traditional practices.

A critical role for a teacher librarian is emerging to include support of students' research and project work, and in more specific terms, leading them to develop and enhance a complex set of 21st century skills increasingly known as the "New Literacies." These new literacies include,

---

[1] Churchill, D. (2017). *Digital Resources for Learning*. Singapore: Springer.

Information Literacy, which has already been recognized as a standard focus of a contemporary teacher librarian in their intervention with school students. In addition, attention must be given to other related literacies such as Critical Literacy, Media Literacy, Digital Literacy and even extending traditional language skills of reading, writing, speaking and listening, and incorporating new skills of Viewing and Representing. Furthermore, traditional library environments are changing, recognizing importance of ambience in stimulating development of new literacies and achieving of learning outcomes. Thus, libraries are increasingly embracing concepts of environments popular with new generation of students (e.g., Starbucks) and successful models of learning commons piloted across the world. Ambience, lighting, decorations, furniture, technologies, etc., all play important roles in the design of modern library spaces. However, roles of a school library are not bounded to the physical spaces, and numerous success stories are telling us that a virtual environment play equally important role. Nevertheless, critical in building a successful school library starts with teacher libraries, whose roles, technology skills, understanding of learning-centered pedagogy, understanding of teacher thinking, and understanding of students and their learning practices, is critical. These critical areas are explored in this book successfully.

Finally, I wish to congratulate the authors on this project, and wish successful impact across the field of education. I am certain that riders will benefit from insights that this book offers. One important aspect of a contemporary publication should by to look forward to the future, and *Effective School Librarianship: Successful Professional Practices from Librarians around the World* is doing that in a pragmatic and an effective way.

Dr. Daniel Churchill
Faculty of Education
The University of Hong Kong

FOREWORD

This book is one of the best platforms currently available for professional sharing among school librarians on an international scale, since it contains alternative points of view from school librarians, who are practicing in different parts of the world.

School librarians are meant to be facilitators/leaders/guides of active, critically reflective learning, as well as knowledge utilization. The value of this book lies in the fact that it provides the readers an overview of many up-to-date practices, which are carried out by school librarians working in a great variety of sociocultural, political, and educational contexts.

This book is not meant to be an operational manual for school librarians. Instead, this book is a collection of inspiring stories told by highly motivated and competent school librarians. The core message of this book is—with a little creativity, imagination, innovation via joint forces, even with very tight budget and resources, we school librarians could do so much more for our children's future and our own.

Daisuke Okada
(Assistant Professor,
Yasuda Women's University, Japan)

FOREWORD

A career in school librarianship is gradually being recognized as a promising profession in Nigeria. Like in other places, the problem with school librarianship in Nigeria is that of recognition and remuneration while colleagues in tertiary institutions (universities, polytechnics, and colleges of education) are more recognized. There seems to be no path for progression for these school librarians' careers. This is because librarians working in the academic sector are recognized by law, and they are allowed top positions in their field, whereas schools are being neglected. However, efforts by the Centre for Educational Media Resource Studies (CEMRS, formerly the Abadina Media Resource Centre) University of Ibadan, Nigeria and the Nigerian School Library Association are gradually paying off.

Students from CEMRS who are working in school libraries such as Dumebi (who was interviewed in this book) and Bimpe (whose picture is on the cover page) are school librarians who have been trained to love what they are doing despite the challenges of getting adequate learning resources for their libraries. With efforts made by the Nigerian School Library Association, school librarians in Nigeria are encouraged to love what they are doing—especially with the eagerness in children who love to read and learn using their school libraries. Few schools in Nigeria—especially privately owned schools—have a library while majority of government owned schools do not! The common denominator for most of these libraries is the lack of adequate resources, but there are school librarians who love what they are doing and work hard in the midst of poverty!

Dr. Fadekemi Oyewusi
(Centre for Educational Media Resource Studies,
Faculty of Education, University of Ibadan)

PREFACE

According to Lynn Barrett (2010, p. 139):

> Librarians need to be trained to become school librarians, well versed in pedagogy and curricula. Only by developing an expertise in the educational arena will they be able to collaborate successfully with teachers, be valued as leaders in their schools and fulfill their potential to contribute to the academic success of their students. Teachers need to receive training about the skills of information literacy and the techniques of effective inquiry learning, where students are challenged to engage with the glut of sources available to them, and to question, select, analyze and synthesize until they are able to discern paths to new understandings and knowledge construction.[1]

All children have the right to quality education and information. The school library plays an important role in this regard. The school library is a flexible, versatile, as well as multifunctional arena that could be used for supporting numerous types of learning, cultural, recreational, as well as leisure activities. It usually consists of the physical library with a collection of traditional printed publications (usually books and periodicals), audio–visual, multimedia materials, courseware, etc. In fact, the resources of a school library are essential for extending the learning experience of students beyond the immediate learning environment, such as the physical classroom and limited class time. For example, in addition to supporting the teaching and learning of the core curriculum, the school library may also provide extra resources to support students in their home reading, along with a wide range of other extracurricular activities for recreational and leisure purposes. As pointed out by Marquardt and Oberg (2011), "In so many diverse ways, the school library offers a wonderful 'bridge' between the school and the knowledge infrastructures outside the school!"[2]

---

[1] Barrett, L. Effective School Libraries: Evidence of Impact on Student Achievement. *School Librarian.* 2010, *58*(3), 136–139.

[2] Marquardt, L.; Oberg, D. In *Global Perspectives on School Libraries Projects and Practices;* Marquardt, L., Oberg, D., Eds.; Berlin: De Gruyter Saur, 2011, p 332.

Joyce Valenza's 2008 metaphor of the school library as a kitchen—a collaborative center for creating, using, and sharing resources—rather than being merely a grocery store, where resources are gathered and taken elsewhere for use.[3]

Education has become more important than ever because information explosion on the Internet and the digital environment simply demands that individuals develop competencies for lifelong learning and grow professionally and be employable. The 21st century conception of learning is about much more than adopting new skills and integrating them into the curriculum or purchasing new technologies and planning them in classrooms—it is the fundamental shift from a teacher-centered learning environment to a student-centered one (Zmuda, 2009).[4] In such a context, school libraries must have flexibility and personalization at the core of services, bringing literacy opportunities and information literacy strategies and activities together by embedding them in multi-modal projects. Information literacy has become a foundational discovery activity that shows students how to investigate and walk through data with wisdom. According to Rossaroli (2011, p. 212), "School librarians also serve as mediators who can guide the teachers and students so that new resources and mediums are integrated with the quality standards of the curriculum. The library also provides a wide range of other services: helping students become critical and efficient information users, promoting access to knowledge and to healthy entertainment, and helping teachers to implement reading plans and to use a wide range of pedagogic strategies."[5]

The school librarian does not only lend books, they give extra value to the contents of the library through different kinds of supports: they organize, guide, and segment resources, and they create a real library without walls, spreading information, informing users, teaching multimedia to users, executing the flow and generation of digital contents, and starting the use of Web 2.0 tools, especially Wikis. (Rossaroli, 2011)

---

[3] Valenza, J. *Library as Domestic Metaphor*. Available at: http://blogs.slj.com/neverendingsearch/2008/08/25/library-as-domestic-metaphor/ (accessed Aug 25, 2009)

[4] Zumda, A. Take The Plunge into A 21st-Century Conception of Learning. *School Library Monthly*. **2009**, *26*(3), 16.

[5] Rossaroli, A. E. The Belgrano Day School Model Project: Using Mobile Technology in A School Library in Argentina. In *Global Perspectives on School Libraries Projects and Practices*; Marquardt, L., Oberg, D., Eds.; Berlin: De Gruyter Saur, 2011; pp 212–222.

With reference to information seeking from students' perspective, students need advice and direct guidance throughout the process of searching, locating, evaluating, reading, and documenting information sources (Kuhlthau et al., 2008).[6] In this context, school librarians take the responsibility for equipping students with essential 21st century learning skills they require to succeed using standards, such as the *Standards for the 21st Century Learner.*[7] With the information explosion over the Internet and its overwhelming impacts on the publishing industry, the school library also plays a vital part in addressing the issues concerning information literacy skills and the use of multimedia. According to Rossaroli (2011, p. 213)[8], "Five centuries separate us from the invention of the printing press, and still the challenge to create content is at a turning point. The ways to access content are redefined and projected to carry on the literacy process and to create a readers' society. On a daily basis, we have new technological challenges and new publishing models that the school library needs to evaluate. Projectors, interactive boards, electronic readers, wireless equipment are the tools that can be used to educate, but we should not forget that the tools are not enough. They must be accompanied by mediators that can guide the teachers and pupils so that the quality standards of the curriculum are blended with the new resources and mediums. In this context, the school library is the fundamental tool in creating and shaping readers...."

Under the leadership of a seasoned, motivated, and committed school librarian, the school library has the potentials to enrich the knowledge among the students in a much larger context that could undoubtedly stimulate boundless thoughts and creative ideas—and hereby enable them to develop skills and attributes for lifelong learning in a systematic manner. Such competencies are unquestionably indispensable for success in the modern "knowledge-driven society."

---

[6] Kuhlthau, C. C., Heinstrom, J., Todd, R. J. The 'information search process' revisited: Is the model still useful? *Information Research.* **2008**, *13*(4). Available at: http://www.informationr.net/ir/13-4/paper355.html.

[7] American Association of School Librarians. *Standards for The 21st-Century Learner.* Available at: http://www.ala.org/aasl/standards/learning.

[8] Rossaroli, A. E. The Belgrano Day School Model Project: Using Mobile Technology in A School Library in Argentina. In *Global Perspectives on School Libraries Projects and Practices;* Marquardt, L., Oberg, D., Eds.; Berlin: De Gruyter Saur, 2011; pp 212–222.

Although school libraries flourished during the 1980s, in a majority of schools, school librarians were absorbed back into regular classroom teaching for the purpose of saving manpower and resources. Meanwhile, school library collections languished and library programs were slowly abandoned. In many cases, only supporting or clerical staff were allocated to maintain most of the remaining resources and other activities associated with the school library. According to Lo et al. (2014, p. 51), "Teacher (or school) librarians are not only managers of the school libraries, they are also educators, administrators, teaching consultants, information specialists, and information literacy teachers."[9] Unfortunately, many novice teachers do not have a clear understanding of the potential contributions of the school library programs to students' overall development process and their academic achievements, as well as their contributions to their academic achievements and overall inquiry-based learning as a whole (Lo et al., 2014).[10] Eventually, this unfortunate situation has led to an absence of a clear and mutual understanding (between school librarians and school management) of the meaning of "school librarianship" and the inherent pedagogical role of school librarians to be completely overlooked. Such a lack of understanding is highlighted particularly in countries where education systems are extremely exam-oriented. Consequently, even in many cosmopolitan cities in Asia, this "can-live-without-a-full-time-school-librarian" situation continued to persist and intensify during the last two decades (Lo et al., 2014).[11] For obvious reasons, promoting a better understanding of the roles and the full pedagogical potentials of the school librarian is undeniably the basis for the advocacy of school libraries in the educational context worldwide. According to Das (2011, p. 287), "Advocacy is a complex and time-consuming undertaking. Advocating school libraries in a rapid changing educational and technological era is challenging but also rewarding. The unique selling points of a library in the educational environment are the basics for the advocating process."

---

[9] Lo, Patrick, et al. Attitudes And Self-Perceptions of School Librarians in Relations To Their Professional Practices: A Comparative Study between Hong Kong, Shanghai, South Korea, Taipei, and Japan. *School Libraries Worldwide*. **2014**, *20*(1), 51.

[10] Ibid, 51–69.

[11] Ibid, 51–69.

AIMS OF THIS BOOK

According to Dr. Helen Boelens[12],

> The majority of school librarians whom I have met at international conferences are very involved in the way school librarianship, and training works *in their own country*. They find it very difficult to look "outside the box"—they believe that the system practiced in their own country is the best. In my opinion, school librarians throughout the world need to take into account the many different aspects of school librarianship throughout the world, in other countries. There are many different forms of "international school librarianship" in context, definitions, etc.—nevertheless their varying methods may be successful for the schools, teachers and students who work with them. During my term as Chair of the IASL (International Association of School Librarianship) research SIG, the question of definitions and guidelines was raised. These new guidelines are very important in defining what a school library actually is, in many different circumstances, from country to country (developing, emerging or developed). The publication of the IFLA School Library Guidelines, 2nd edition, which were published in June 2015 (http://www.ifla.org/files/assets/school-libraries-resource-centers/publications/ifla-school-library-guidelines.pdf). This document needs to be publicized.

This book contains chapters of interviews with individual school librarians discussing their practices, challenges, as well as a variety of topics related to their professional practices. Using this book as a professional platform, these practicing school librarians from different parts of the world are sharing their knowledge and experiences for a wider global audience. By gaining insights into their perceptions, there was impetus to consider ways to guide changes aimed at aligning the professional school librarian practices with the "actual," as well as the "preferred" learning environment in different geographical, national, and cultural contexts.

This book aims at bringing together topics, methodologies, approaches, and experiences of school librarians practicing in different parts of the world, and the goals could be summarized as follows:

1.   To provide a global perspective on projects and practices related to school librarianship—thereby:

---

[12] Dr. Helen Boelens, Ph.D., School library researcher and consultant, Former Chair, IASL Research SIG, The Netherlands.

- Inspiring as well as fostering professional developments for the overall school library enhancement at different levels;
- Breaking new grounds in the research literature related to school librarianship;

2. To carry out in-depth studies on the impacts of school libraries in the global context on students' learning preferences—with the aim of shedding new insights on youth behavior toward modern technology and the actual implementation in schools.

3. To update practicing school librarians and classroom/subject teachers on the progress, nature, evolution, pedagogical potentials, as well as possible impacts of school libraries in a variety of learning environments under different cultural contexts.

4. To call attention to school library research that is written from the perspective of direct, in-depth and hands-on experience among practicing school librarians—thereby suggesting new strategic directions and feedbacks for consideration of how to improve the learning environment—hopefully to make a significant positive impact on literacy and curriculum support.

5. To make sense of how librarians make the most of the learning environment in a variety of school situations, and thereby contributing to an understanding of useful strategies to further strengthen the role of school libraries in the school community.

6. To examine the collaborative relationships between classroom teachers and school librarians, as well as to provide real-life examples on how such collaboration could impact on the teaching and learning for the school community as a whole, particularly in relation to the design and planning of instructional units across different subject disciplines.

7. To demonstrate how school librarians at school libraries in different remote communities design and implement programs meant to promote literacy among their students.

8. To present projects and practices addressing the challenges of supporting the developments in the different areas, that is including contexts where many children do not have access to formal education or reading materials on a regular basis, for example:
   - Motivation to read voluntarily for pleasure and for information
   - Basic information literacy skills for the navigation, evaluation and use of information

- Competence as independent learners—a key factor for successful inquiry-based learning

CHOICE OF METHOD

Qualitative interviews were used for illustrating the underlying reasons for individual interviewees' professional practices and actions, as well as decisions. Interview techniques enabled participants' attitudes and opinions to be fully expressed in their own words, and allowed space for a variety of, sometimes contradictory, points of view to be aired. In addition to allowing individual school librarians to freely discuss their perspectives and experiences, such natural and free conversational interviews also enabled maximum flexibility for more open, spontaneous, and instant exchanges of ideas without any preconceived expectations on my side. Not only does the interview approach provide opportunities for open discussions where both participants and researcher can "construct meaning" together, it is also "essential for the understanding of how participants view their world."[13] The insights gained through exploration with individual school librarians of viewpoints are also important for gaining the broadest possible perspective of the issues addressed. In other words, the qualitative approach allows us and the readers to extend the depth and breadth of understanding the nuances of differences between learning environments in a variety of cultural contexts. Comparisons of perceptions between practicing school librarians appearing in this book also provide useful feedback in the context of a learning environment in the digital age and globalized world.

INTENDED READERS

The individual school librarians presented in this book are eager and enthusiastic to learn from one another, and to strengthen their views on their profession—offering innovative ideas and sound techniques for professional developments among school librarians. This book will be a

---

[13] Rossman, G.B., Rallis, S. F. *Learning in The Field: An Introduction to Qualitative Research.* Thousand Oaks, Sage Publications, California. p 124.

valuable and practical addition to many libraries and personal collection of educators—serving as a useful reference for classroom teachers, librarians, and school administrators who wish to strengthen their professional practices in the area of information literacy and learning and to increase goal achievement among their students.

This publication is also of interest to classroom teachers in any subject disciplines and practicing librarians in particular who are trying to increase their knowledge and skills in school librarianship. It is hoped that this book would contribute to a better understanding of useful strategies to further strengthen the roles and practices of school librarians in the education communities on a global level.

ACKNOWLEDGEMENTS

This is our fourth book on librarianship, but very first joint book project that is dedicated to documenting the professional lives and practices of school librarians. The original idea of this book was built on our earlier empirical study, *Attitudes and Self-perceptions of School Librarians in Relations to Their Professional Practices: A Cross-National Comparative Study*—set out to examine, explore, and compare how school librarians in different education systems perceive their own status within the school community by looking at their relationships with their principals and other classroom teachers.

From the very first idea to its final publication, this interview book project has accompanied us for several years. We are indebted to a number of people who have generously supported us in writing this book. To begin with, this book would never have been materialized without Dr. Helen Boelens. With her years of research experiences and incredible connections in the field of school librarianship, we were able to get interview after interview from some of the most confident and competent school librarians practicing throughout the globe sharing the secrets behind their successful, and yet meaningful stories.

Throughout the course of this interview book, Heather Rogers, Dickson Chiu, and I had the remarkable opportunity to converse candidly with 36 school librarians practicing in different parts of the world, including Africa, Asia, Australia, Europe, North and South America, and so on. To acknowledge their immeasurable contributions, we would like to extend our most sincere gratitude and appreciation to all 36 school librarians who took part in this interview book project—taking their time out of their hectic work schedules, overcoming language barriers, to share and exchange their valuable professional experiences and insights with us.

We reserve a special note of thanks for Apple Academic Press and, in particular, our editor, Sandra Sickels, for having faith in us by taking up this book project.

Last but not least, we would like to express our gratitude to Prof. Chutima Sacchanand (President of the Thai Library Association)—who not only helped us expand the coverage and content of this book by looking for highly motivated school librarians in Thailand who were prepared to take part in project, but has also assisted with the translation of interviews from Thai into English.

School librarians are not only managers of the school libraries' daily operations, they are also educators, administrators, teaching consultants, facilitators, servant leaders, homework, and student project coordinators, and much more. They are without doubt "unsung heroes" of information literacy, pedagogy, voluntary reading, curriculum development, student homework and project coordination, classroom-library partnership, and extracurricular/recreational collaboration—who deserve every single bit of recognition like other classroom teachers, and should unquestionably be considered an integral component to any educational system, regardless of country or region.

—Dr. Patrick Lo and Dr. Dickson K.W. Chiu

I would like to express my gratitude to those who provided support and encouragement throughout this process. I would first like to thank Dr. Lo and Dr. Chiu for the opportunity to get involved in this project and for the chance to learn more about the wonderful profession of school librarianship. I am grateful to Judy Ashby for her encouragement as my own school librarian, mentor, and friend. I want to extend thanks to Annie Lyon for her guidance and advice that helped me enter the field of librarianship. I would also like to thank Mami Kobayashi for her insight into Japanese librarianship and her kindness during my time in Japan. Finally, I would like to thank Zachary Alapi for his continued support and encouragement.

—Heather Rogers

ABOUT THE AUTHORS

DR. PATRICK LO

Dr. Patrick Lo is currently serving as Associate Professor at the Faculty of Library, Information & Media Science, University of Tsukuba in Japan. He earned his Doctor of Education (EdD) from the University of Bristol (U.K.), and has a Master of Arts (MA) in Design Management from the Hong Kong Polytechnic University, a Master of Library & Information Science (MLIS) from McGill University (Canada), and a Bachelor of Fine Arts (BFA) from Mount Allison University (Canada). He also took part in a 1-year academic exchange at the University of Tübingen in Germany from 1990–1991. He is proficient in Chinese (both Cantonese and Putonghua), English, and German.

Dr. Patrick Lo has presented about 100 research papers and project reports focusing on librarianship, humanities, and education at different local and international workgroup meetings, seminars, conferences, etc., including: Mainland China, Hong Kong, Austria, France, Germany, Italy, Japan, Korea, Turkey, United States, and Sweden, and at institutions including the Library of Congress (U.S.), Austrian National Library (Vienna), University of Vienna, National Library of France (Paris), National Institute of Informatics (Japan), Konrad-Zuse-Centre for Information Technology (Berlin), etc.

His research interests and areas of specialty include: comparative studies in LIS; and art and design librarianship and information literacy. His recent publications on LIS include:

Conversations with the World's Leading Opera and Orchestra Librarians. Lanham, MD.: Rowan and Littlefield. (2016).

Preserving Local Documentary Heritage: Conversations with Special Library Managers and Archivists in Hong Kong. Hong Kong: City University of Hong Kong Press (2015).

"Comparative study on M-learning usage among LIS students from Hong Kong, Japan and Taiwan." *The Journal of Academic Librarianship* (2015).

"Enhanced and changing roles of school librarians under the digital age." *New Library World* (2015).

"Information for inspiration: Understanding information-seeking behaviour and library usage of students at the Hong Kong Design Institute." *Australian Academic & Research Libraries* (2015).

"Why librarianship? A comparative study between University of Tsukuba, University of Hong Kong, University of British Columbia and Shanghai University." *Australian Academic & Research Libraries* (2015).

Patrick is currently working on another interview book project entitled: *Conversations with the World's Leading East Asian Librarians, Archivists and Museum Curators.* This book will include interviews with East Asian librarian form the Vatican Library, State Library Berlin, Bavarian State Library, National Library of France, East Asian Library at Princeton University, Hong Kong Chinese Martial Arts Living Archive, and others.

HEATHER ROGERS

Heather Rogers is currently a Master of Information Studies (MISt) candidate at McGill University focusing on librarianship and information literacy. Previously, she was an assistant English language teacher for the Japan Exchange and Teaching (JET) Program in Fukushima, Japan from 2013 to 2016. She graduated from The American University in Washington, D.C. with a Bachelor of Arts (BA) in International Studies and Japanese.

DR. DICKSON K.W. CHIU

Dickson K.W. Chiu received the B.Sc. (Hons.) degree in Computer Studies from the University of Hong Kong in 1987. He received the M.Sc. (1994) and the Ph.D. (2000) degrees in Computer Science from the Hong Kong University of Science and Technology (HKUST). He started his own computer company while studying part time. He is now teaching at the University of Hong Kong and has also taught at several universities in

Hong Kong. His research interest is in library and information management with a cross-disciplinary approach, involving workflows, software engineering, information technologies, management, security, and databases. The results have been widely published in over 200 papers in international journals and conference proceedings (most of them have been indexed by SCI, SCI-E, EI, and SSCI), including many practical master and undergraduate project results. He received a best paper award in the *37th Hawaii International Conference on System Sciences* in 2004. He is the founding editor-in-chief of the *International Journal on Systems and Service-Oriented Engineering* and the EAI Endorsed Transaction on e-Business. He also serves on the editorial boards of several international journals. He co-founded several international workshops and co-edited several journal special issues. He also served as a programme committee member for over 200 international conferences and workshops. He is a Senior Member of both the ACM and the IEEE, and a life member of the Hong Kong Computer Society.

PART I

Africa

CHAPTER 1

A SCHOOL LIBRARY IN A PLACE WHERE THERE HAS NEVER BEEN ONE BEFORE! BUILDING A READING CULTURE IN TO THE DEMOCRATIC REPUBLIC OF CONGO

KATHERINE SHAW

General Director, Academie Bilingue du Congo (ABC), Blvd Nyamwisi, Beni, Nord Kivu, Democratic Republic of the Congo (DRC)

Please provide a self-introduction and tell us about your professional and educational backgrounds. What did you study at university? Are you a second-career school librarian—meaning that did you have other careers before becoming a school librarian?

My name is Katherine Shaw, and I was educated around the world, including primary schools in the USA, France, and Burundi, and secondary school at an American system boarding school in Kenya. I completed my Bachelor's degree in Nursing at Calvin College[1] in Michigan, USA. I worked for eight years as a nurse in a variety of settings (inpatient hospital, urgent care, maternal and infant health) before moving to the Democratic Republic of Congo (DRC). I began to work informally as a school librarian after living and working in Beni, North Kivu region in the DRC.

Your previous international experiences (in France and Burundi), and your previous professional as a nurse—do they in any way contribute to your current work as a school librarian in Congo?

My experiences in France and Burundi allowed me to learn the French language, and this also exposed me to French teaching styles and curricula,

---

[1] Calvin College—Homepage. Available at: http://www.calvin.edu/.

as well as a broader range of books—in general, helping me to understand the complexity of providing appropriate literature for children in a variety of cultures and settings. My profession as a nurse gives me a deep understanding of the need for lifelong learning the importance of reading continually to keep my knowledge base up-to-date and relevant—something that is not often possible for children or adults in Eastern Congo due to a lack of books and access to digital resources.

What is the average literacy rate amongst population in your region of Congo?

Because the last census was taken in 1980, statistics of any kind in our region are certainly unreliable at best. UNICEF cited countrywide literacy rates of 78.9 and 53.3 % for males and females, respectively, aged 15 to 24, but the Northeast region where we work falls continually behind in nearly all respects. (You can find the UNICEF statistics here http://www.unicef.org/infobycountry/drcongo_statistics.html)—there are some interesting numbers further down about number of books in a household.

In your region of Congo, if a majority of the high school graduates would only end up working in factories, mines, or doing laborious farming work—why would they still need to have a high level of reading skills in English? Additionally, how would these high school graduates benefit from it?

At our school (Academie Bilingue du Congo or ABC), we have a majority of students who actually come from relatively affluent backgrounds, and who may not work as laborers. Our goal is also to foster new leaders for the country who will defy the odds and become reliable, ethical, and competent leaders, as well as lifelong learners, and so forth. I think assuming that people do not "need" to learn something, whether it is English or something different, has a way of preventing them or their children from future success. I believe desire and thirst for knowledge and understanding are forerunners to success and believing in yourself, and if we tell people not to bother learning skills they will not "need." We are putting an unfair tether on their ability to grow and dream as individuals and communities.

Could you describe the Internet infrastructure in Congo? If the Internet network is very much developed, would it be more effective to bring iPad or digital tablets to the students—thereby enabling them to have easy access to large amount of reading materials via the Internet, instead of investing a great deal of manpower and resources to build a physical library on school campus?

We currently have no Internet access at our school, and consistent access to Internet is not available in town. There is no power grid in Beni and people who wish to access Internet go to small Internet cafes and pay by the minute. In the past year or so, mobile Internet has become available through cell phone companies, but only very wealthy people can afford a phone that will allow them to access it. It remains very expensive as well, costing $10 for one 1 GB of Internet, so downloading large amounts of resources is very costly. In short, obtaining technological materials and the Internet will unfortunately also require a great deal of manpower and resources. Even so, we want to bring Internet access to the school as soon as possible, as we believe access to technology and the wealth of written resources online is necessary to give them a relevant and well-rounded education. Even so, a physical library with beautiful books to touch, share, bring home, and read as a group provide learning opportunities (for the entire family—some of our student's parents have begun reading because their children bring home books) that technology alone cannot address.

Is the Congolese curriculum very much exam-based or on the contrary inquiry-learning-based? If the current elementary and secondary curricula in Congo are only meant to increase the overall literacy rate (and to eliminate illiteracy) amongst the general population—what roles do the school library and school librarians play in this context?

The Congolese curriculum is exam-based and leaves very little room for inquiry or critical thinking, placing a great deal of emphasis and time on rote learning—mostly memorization and copying. I think a library will play a crucial role in allowing teachers and children to delve deeper into ideas that may remain very superficial in the curriculum, and to take them instead as opportunities for more reading, discussion, and new challenges. Also, we do not want our students just to know how to read, but we want them to use reading as a means to greater understanding and learning, and an avenue to explore interests and ideas that are not presented in the standard curriculum.

Could you describe the social backgrounds of your students? What is the average income per household in your region of Congo? What kinds of jobs do a majority of their parents do?

Our students mostly come from relatively privileged homes, although some are not. The CIA World Factbook data states the average annual income in DRC is $ 400, but again our region of North Kivu tends to fall in a lower socioeconomic position than many other parts of Congo. In our region nearly everyone tries to make ends meet in a variety of ways since there are few stable jobs, and nearly everyone farms either small or large plots to help ensure the family has food and hopefully a little extra money from selling extra produce (some have been unable to do this in recent months due to violence and kidnapping in the farming regions outside of town). Those who are able often seek to supplement that income through some entrepreneurial endeavor such as raising and selling animals, making handicrafts, doing hair/nails, running a small shop, etc. Some parents of students at our school are in higher socioeconomic brackets and have stable jobs (for instance, working in the coffee trade, at a local bank, own a shop or restaurant in town, working for an NGO or the UN). Most children have two parents, multiple siblings, and often, extended family members or close friends living together with them in one house. Many families house refugees from neighboring regions where violence has occurred. Some have lost family members to illness or violence due to the instability in our region.

Choosing a career in school librarianship, was it an active choice out of personal interest? Or it was by chance and circumstance?

Upon arrival in DRC, my husband and I were fully committed to working alongside a locally-led NGO called Congo Initiative, whose flagship project is the Christian Bilingual University of Congo (known as UCBC, its French acronym).[2] The University, whose vision is to create transformative leaders to change the future of Congo, asked me to champion the

---

[2] Christian Bilingual University of Congo—Homepage. Available at: http://www.ucbc.org/.

start-up of a much needed primary school to serve the region. In addition to serving a variety of roles at the school (including general director, classroom teacher, curriculum development) one of my early priorities was to establish some form of school library for our students. There is currently no other library, or source for children's books of any sort in Beni, or within reach of children and families in the town and environs of Beni. So, it felt like a necessity, driven by circumstance as well as personal interest.

Are you currently working as a solo librarian in the whole school?

Managing and using the library has been a team effort among the teachers, but I have and continue to lead the project more than anyone else. The library and its programs are still not well-organized due to the many roles all of us are playing.

"Managing and using the library has been a team effort among the teachers"—could you tell us in details why you need to team up with other teachers? How do you share workload and responsibilities when it comes to teaming up with other teachers? Have you encountered any difficulties, challenges, and reluctance amongst other teachers in this regard?

On Fridays when children take home a book, classroom teachers are individually in charge of tracking and logging what books come home, and whether or not they are returned. Another teacher does the longer-term tracking of the catalog and ensures that books are returned/repaired each term. Mostly the difficulty lies in all of us having so many other responsibilities—it is hard to find time to do much long-term planning or thorough organization because we are all doing many other things. Teachers are willing, but it is hard to have time.

Could you describe your typical day at work as a school librarian?

Each of our primary classes has time in the library each day, during which classroom teachers are assisting them with finding books and also conducting read-alouds or giving brief lessons related to library use or books. Children are allowed to bring a book home on the weekends, therefore, on Fridays teachers help children select a book and then log it in a simple form to track which books go home.

Why are read-alouds considered important and useful for your students?

Students are just learning both French and English, so read-alouds allow them to listen to proper pronunciation, flow, and speech patterns in new languages. It is a great opportunity to engage students with text in creative ways (sometimes we have props, act out the stories together, or do other reinforcement activities after reading) and to work on pre- and early reading skills like letter recognition, rhyming, context clues, comprehension, and so forth. For kids who have never read books at home, simply learning to listen well and track what is happening in a story is a new skill, and read-alouds are a great way to encourage an early love of reading and books.

The official language of DRC is French—what advantages would it gain by investing so much manpower and resources into building a school library that would help develop children's reading skills in English?

While French is an important global language, English continues to grow in its place as the global language, and a majority of published literature, research, and online content is in the English language. As such, school administrators, and also parents/community members in DRC recognize that English reading skills are a crucial component if they want their children to be able to engage academically at a global level.

Do you need to take up any classroom teaching duties, in addition to fulfilling your roles as a school librarian?

Yes, we currently cannot afford a full-time school librarian, and all of us are serving in multiple roles.

As a school librarian in your region—is there a nationwide or region-wide syllabus or curriculum that you need to follow, in terms of performing your work as a school librarian? If not, do you think it is feasible to implement a region-wide syllabus for school librarians? The absence of such a syllabus—do you think it is an advantage or disadvantage?

There are no other primary school librarians that I know of in our region. Historically, books have not been available for routine use or for outside reading in the Congolese curriculum, so there is no nation or region-wide syllabus. I would love to have such a resource for us to use as we try to set up and get running as a library, and hopefully in the future as we try to open a library that neighborhood children can also access.

What are the expectations amongst your students, other classroom teachers and the senior management in the school library, and in you—in the context of supporting the overall learning and teaching, as well as the development of other recreational activities of the whole school?

There are no expectations; in fact, having access to books for personal research and use is something that almost all teachers and students had never experienced until we started our school. Teachers initially expressed that they did not really know how to use resources and found it over-whelming, but quickly developed a desire to read and research more to help with lesson preparation, and children look forward to library time and free reading time each day. I would like to put a more strategic curriculum in place to better support teaching and learning through library activities, but we have not had the resources to do so yet.

"Teachers initially expressed that they did not really know how to use resources and found it overwhelming"—could you give examples to illus-trate your point here?

In the beginning of the year, I would express that teachers should use books to find lesson ideas, images, or to find/confirm information they were going to teach, but since most teachers had little or no experience with either libraries or books in general, it was hard for them to know where to start. They would continue to teach from memory, as they did previously. I had an interesting situation where a first-grade teacher began to teach a geography unit about the continents. We had a student come in and tell her teacher that her dad was really upset that her "teacher was telling her lies" because he believed there were only five continents—that was what he was taught in school. His daughter protested to him that she has seen the seven continents on a map in a book. He actually came to the school to prove his point, and the teacher was able to show him the world map in our library, as well as an atlas on the shelf and her curriculum book to back up her teaching. That was a great real-life example of why it is important to understand where information comes from, and why reliable sources matter.

If the functions and purpose of the school library is only meant to provide basic reading materials for the students, why would you still need or prefer a school librarian to manage the library or to take the functions of the school library to the next level? Could a regular clerical staff with high-school level education, with no training in school librarianship to oversee the basic daily operations of the library instead? Would it be more cost-effective this way?

Currently, we are only able to provide basic reading materials for students and families, and untrained staff is able to keep a log of books. However, having been privileged to use many libraries in my lifetime, I know how transformative a really excellent and engaging library can be. For instance, I am residing in Ann Arbor, Michigan for a few months, and I am amazed to see the variety of activities and services that the city library provides in addition to its excellent book selection and technological capacity—there are musical instruments, sewing machines, cameras, and other creative materials that can be checked out, not to mention the many opportunities to attend lectures, live performances, an award-winning story-teller on staff, and children's activities of all types. I doubt that these are cost-effective activities, and yet, we largely take them for granted in the United States and much of the western world. When I look at the children I am privileged to live and work with in Beni DRC, and I feel they are every bit as deserving of such opportunities as my own children. Just providing books and a couple of used computers is certainly a small and cost-effective start, but I do dream of pouring much effort and many resources into providing something better for our community there.

Please give a list of successful library programs (supporting students' overall learning and teaching of other teaching staff) initiated by you as a school librarian?

I do not have any named programs, but I conducted seminars with staff teaching them how to use research books, how to incorporate books into classroom activities and lessons, and types of questions to ask after free reading time to help children make the most of unstructured reading. We also have weekly all-school story times (we are currently very small, with K-2 ages, and 34 students total) where we try to engage reading material in fun and hands-on ways.

What are the major challenges and difficulties faced by you as a school librarian?

Lack of resources, knowledge/training, technology, sufficient time for good management and planning.

Which parts of your job as a school librarian did you find most rewarding?

Seeing children and teachers engage with books and get excited about reading and discovering new things on their own. This is truly the first place many of them have ever had the ability to engage with written material of their own choice, and it is a joy to see that happen.

The professional knowledge, skills, roles, and other job-related competencies for a school librarian—have they undergone major changes in your region in the last 5–10 years? In your opinion, what is the future for school librarians in your region?

As previously explained, primary school libraries, and primary school librarians do not exist in our part of DRC. To me, that means the future is bright—imagine what an impact even one small project like ours can have! I envision a future of many school libraries and librarians, sharing knowledge, and resources with children and populations who had no access to such information previously. In spite of many challenges, I am very hopeful!

Having a passion for school library work, and do the kinds of work that you are currently doing in Congo—do you think it is something that is inborn (some people would say it a calling) or it is something that could be developed over successful experience and exposure?

I think experience and calling worked together to draw me to this work. I do not know if I would have felt the sense of calling and desire to do this work at the school and library if I had not seen firsthand the need that is present, and loved dearly the people around me, noticing the difference between many of my friends' opportunities and my own. I think others interested in such work need to take time first to get to know and love a community deeply, either directly or through a locally embedded connection. I think that experience and exposure (which may in turn develop into a feeling of calling) are crucial to develop a meaningful and lasting institution.

As a school librarian, how would you go about to acquire books and other materials to build the school library collection—via purchasing from book vendors or via donations?

So far, because our funds are very small and currently, all are needed just to keep the school running, we have relied entirely on donations to obtain books. Most book donations have been random, but we have had the opportunity to select some. In fact, we currently have a friend raising some money to send some books to us and a few other schools in Eastern Congo. If you know of anyone who might be interested in supporting this initiative you can find more information at this website: https://www.razoo.com/us/story/books-for-eastern-congo.

If they were to lay off the school librarian or to close down the school library completely, what kind of impact do you think it would have on the overall learning and recreational needs of the whole school community?

The children, teachers, and families would be extremely saddened. I think it is currently one of the most important and standout features of our school, because it allows children to do their own learning and exploring in a culture and educational landscape where there is little room and opportunity for that. Teachers also rely on reference books to prepare their lessons because Internet is not currently available. Children and teachers would not be able to bring books home to share with family members either, which would be very sad for everyone.

What kind of attributes does a motivated and successful school librarian always possess?

A love of reading, words, stories, and learning new information, and a desire to share that with others to help them grow and learn in turn, ability to organize and plan well, and at least the basic technological capacity.

Throughout your career as a school librarian or your career as an educator in Congo, did you ever have any regrets or second thoughts?

I do not know if I would call it regret, but I often feel overwhelmed by the task. There is so much that we need, and very few people to share the burden. At the same time, it is extremely rewarding to work in this capacity, because there is no one else doing it! While I am not the most qualified or best person for the job, I am present and willing, and that means a lot when these kids would otherwise rarely or never get to touch a book themselves.

Do you have any other interesting stories that you would like to share with the readers? Do you have any stories regarding what you do as a school librarian has changed the lives of the students you serve?

I would just say that it gives me great joy to see the transformation in our students from the first day when they enter the classroom and hardly touch the books, to a few weeks later when they spend every free minute reading a book or sharing one with a friend. When children ask me if we have any more books on a particular subject so they can read more, or say they saw something they never knew existed in a book, I know our small project is making a difference!

Katherine Shaw
General Director, Academie Bilingue du Congo (ABC), Blvd Nyamwisi, Beni, Nord Kivu, Democratic Republic of the Congo (DRC)

Preschool Teacher Annie Kahindo teaches preschool children during daily read-aloud time

Early elementary students choosing their favorite books during bi-weekly library time

Preschool students sharing a book in their classroom reading corner

CHAPTER 2

OUR PASSION TO PROVIDE SCHOOL LIBRARIES TO THE CHILDREN OF ZIMBABWE

HOSEA TOKWE

Chief Library Assistant, Midlands State University[1], Library Department, Gweru, Zimbabwe

Please provide a brief self-introduction and tell us about your professional and educational backgrounds. Could you tell me what you studied at university?

My name is Hosea Tokwe, and I am a Chief Library Assistant at Midlands State University Library. I am a fully qualified librarian with a Higher National Diploma in Library and Information Science (LIS).

Could you briefly describe the education system in Zimbabwe? For example, how many years of preschool, how many years of primary and secondary schooling before students can enter university? What is the medium (language) of instruction in Zimbabwe? Are all textbooks written in English? Are all classes conducted in English?

Zimbabwe's education system is the best in Sub-Saharan Africa with a more than 90 % literacy rate. It includes two years of preschool, seven years of primary education, and six years of secondary education. Students who pass Form Six with three subjects in Commercials, Arts, or Sciences with five points or above are eligible to enter university education. Zimbabwe, as a former British colony, has always been using English as a medium of instruction. However, not all of the textbooks are written in English. In both primary and secondary education, lessons are

---

[1] Midlands State University (Gweru, Zimbabwe)—Homepage. Available at: http://ww4.msu.ac.zw.

conducted in English, Shona, and Ndebele, which are the two major indigenous/vernacular languages. Commercials, Arts, and Sciences subjects are conducted in English, whereas the vernacular languages are taught in Shona and Ndebele.

Could you tell me the social and economic backgrounds from which the school students in Zimbabwe come? In addition, what kinds of work do a majority of their parents do for living?

A majority of students come from middle-income home background, and their social life is on average fairly good as they have access to good amenities, can afford to go to school, have access to clean water and good shelter at home where they can study. Generally speaking, parents are in the civil service; others are teachers, nurses, office workers who all earn decent income enough to send students to school.

Choosing a career in school librarianship, was it an active choice out of personal interest? Or was it by chance and circumstance?

I am not a school librarian, but I have great passion for school librarianship. I am an individual member of IASL (International Association of School Librarianship)[2]. I helped to establish the Matenda School Library[3], and have presented a paper on school librarianship in Africa at an international conference.

To work as a school librarian in Zimbabwe, what are the minimal professional qualification requirements?

It depends on the types of school:

- For a government school—one would need either Secondary School Certificate or Certificate in Library and Information Science.
- For private schools—either a college diploma in Library and Information Science (LIS) or a university degree in LIS.

---

[2] International Association of School Librarianship (IASL)—Homepage. Available at: http://www.iasl-online.org.

[3] Matenda School Library Project—Homepage. Available at: https://tokwehosea.wordpress.com/2014/05/23/matenda-school-library-project/.

For the other school librarians working in Zimbabwe, could you describe their typical day at work?

A typical day of a school librarian starts with shelving books in shelves in proper order. This is followed by receiving previously-loaned books and checking whether there are no overdue books. Later on, it is office work—either recording new books in accession register then classifying and cataloging the books; for schools that do not have a computerized system, this will be done manually whereas for those with library computer software, everything is entered into a computer system.

The currently education system in Zimbabwe, is it still very much exam-oriented (with heavy emphasis on rote memorization—that is, focusing more on preparing students to do well on public exams)? Or is it geared more towards inquiry-based learning?

Zimbabwe's education system is exam-oriented in preparation from secondary to tertiary education. Yes, the major focus is for students to do well on public exams; inquiry-based learning happens at tertiary level.

As a school librarian in your region, is there a nationwide or region-wide syllabus or curriculum that you need to follow, in terms of performing your work as a school librarian? If not, do you think it is feasible to implement a region-wide syllabus for school librarians? The absence of such a syllabus—do you think it is an advantage or disadvantage?

Here in Zimbabwe, we are preparing to come up with a School Library Association, and I will take the advocacy role to come up with a Constitution; already, the Hungarian SLA[4] has sent me their Act of the Association, which constitutes the legal framework they will follow. Through my networks and with the fellow Zimbabwean who happens to be the IASL Southern Africa Regional Director, we will be guided by the Constitution in helping us formulate and implement a syllabus that we will present to the Education Ministry. Otherwise, currently, absence of a syllabus puts schools at a disadvantage.

What are the expectations amongst the students in Gweru, other classroom teachers, the senior management in the school and the school librarian—in

---

[4] School Librarian's Association, Hungary—Homepage. Available at: http://www.ktep.hu/english.

the context of supporting the overall learning and teaching, as well as the development of other recreational activities of the whole school?

As a librarian with a passion of school libraries' development through my numerous visits to schools, I have noticed a great interest from school authorities. For example, there is one rural school I have been to whose school administration agreed to establish a school library, and the School Development Community has converted a classroom into a school library building—an ideal school library that now needs books. I will be visiting this school in few days (I look forward to sending you the photos).

Please give a list of successful library programs (supporting students' overall learning and teaching of other teaching staff) initiated by you as a school librarian?

I helped establish the Matenda School Library at Zvishavane here in Zimbabwe through the Matenda School Library Project from 2007 to 2010. I used to make several journeys to this school—since it is a remote rural school, I sometimes had to walk a distance of 25 km on foot—risking my life in deep forests carrying a bag full of books. The Matenda School Library was successfully launched in July 2010, and it continues to receive books from book donors.

What are the major challenges and difficulties faced by you as a school librarian?

From personal experience, here in Zimbabwe, Africa, there are number of difficulties and challenges faced by a school librarian, namely a lack of professional recognitions, poor remuneration (in Rural Schools—run by School Development Committee), inadequate library materials like reference books (e.g., dictionaries, atlases, manuals, almanacs, yearbooks, etc.), unavailability of library tools or resources to help in classification and cataloging of library materials, poor library equipment, insufficient sitting space for readers, and so forth. The great challenge, above all, is the lack of reading materials. It is only through the availability of books that will help transform the lives of the poor rural child in Africa.

Which parts of your job did you find most rewarding?

I am not a school librarian, but as a professional with passion for school libraries. What I find most rewarding is to make available books to poor rural schools, and seeing those poor rural children holding a book and reading to others. I had this memorable experience when I presented a book donation of close to 50 books to the Matenda School some five years back.

The professional knowledge, skills, roles, and other job-related competencies for a school librarian—have they undergone major changes in your region in the last five to ten years? In your opinion, what is the future for school librarians in your region?

In my country, Zimbabwe, there are three library schools (namely at Harare Polytechnic, Bulawayo Polytechnic, and Gweru Polytechnic) where students are taught the skills of running a school library. Yes, the professional knowledge, skills, and roles for school librarians have undergone major changes in that students get practical work experience in universities, colleges, and other institutions during course of their careers. So, the future of the school librarians is bright because also the local Library Association invites school librarians for refresher workshops. Zimbabwe librarians are planning to come up with a (national) school library association, and as someone who has had a first-time personal experiencing after attending the International Association of School Librarianship, I am ready to impart and help foster greater development through formulating a strategic plan that will equate school librarians with those in an international position.

You mentioned you are under the process of formulating a strategic plan for a local school librarian association—could you tell me what are the main Goals and Agendas of your strategic plan?

It is something that is still in its infancy—we are still gathering information from German School Library Association and Hungarian School Library Association by looking at their crafted Bills on School Library Association Act. So, our plans are still in the pipeline, as we have yet to identify the relevant stakeholders so that with the necessary funding we can hold an All-Stakeholders Workshop to brainstorm and come up with a

Draft Plan. So, at the moment, we have yet to identify goals and agendas of the Strategic Plan.

Having a passion for school library work, do you think it is something that is inborn (some people would say it a calling) or it is something that could be developed over experience and exposure?

I believe that it is inborn because good school libraries thrive where there is that human touch of creativity and enticing the user to use the library, to read, acquire lifelong knowledge. With experience and exposure, one only learns to perfect his/her skills in running the school library, but passion brings with it new dimension that lures the readers to come and seek knowledge for educational achievement.

School libraries/school librarians and inquiry-based learning—do you think they go always hand in hand? In the school environment, true inquiry-based learning could not be carried out without a proper school library that is managed by a professionally trained school librarian?

Yes, they go hand in hand, but, unfortunately, in Zimbabwe, we have no well-established school libraries. However, efforts are being made to come with school library and resources centers in some private schools that are well funded. These types of schools can afford to employ professionally-qualified school librarians with either a college diploma or university degree.

School libraries in Zimbabwe with regards to government institutions are in bad shape with outdated collections. In mission and private schools, the situation is better than in public schools. Most government school libraries are run by non-professionals and students. In some cases, they are run by teachers. In mission schools, they are run by qualified librarians with national certificates or better and in some instances they are national-diploma holders. Private Schools employ professionals with ND (national diploma), HND (higher national diploma), and BSc (Bachelor of Science) in Library and Information Science.

Private school libraries have budgets, government schools do not; they rely solely on donations of irrelevant library materials from NGOs or private organizations. Some mission schools have budgets. On the other hand, in private schools, the board of trustees monitor their operations as for government schools—there is no school library service to monitor operations. National Library and Document Service, a statutory body set up by government in 1985, is supposed to monitor government school libraries, but it is dysfunctional. In private schools, the School Development Association is doing great job because every parent donates to the library a new book as and when a request has been made.

Learning environments provide an opportunity to discover and learn new things or ideas, whilst the library space quenches that thirst of inquisitively on part of the student who wants to discover further from what they picked from normal classroom teaching. The library and librarian are equipped with research support and assistance tools, are well versed in ILS. In essence students and teaching staff benefit immensely from the library resources, and librarian's expertise in executing an excellent inquiry-based service support. The school library will always play an equivocal role in enhancing an effective inquiry-based learning.

Could you talk about the School Library March event, which you organize during the International School Library Month?

Well, the International School Library Month is an advocacy program that falls under the International Association of School Librarianship. It is celebrated in the month of October worldwide, in recognition of the importance of school libraries. Here in Gweru, Zimbabwe, I was the first local librarian to successfully organize and celebrate it on the 30th of October, 2010, at CJR primary school. I also successfully organized more celebrations in 2012 to 2015. The beauty of it is that we marched in the streets with a big banner much to the delight of the on-looking public. Over the years, I have had support from the School Library Network (UK) in the form of a small funding to buy t-shirts and refreshments to kids and guests, and I have also had support from the following people: Margaret Ling (UK), Barbara Band (UK), Karen Hans (UK), Dr. Helen Boelens (Netherlands), and Dr. Albert Boerhorst (South Africa) and also, most importantly, the Darien Book Aid International that has been sending me one or two boxes all these years.

During International School Library Month Celebrations, we encourage schools to establish libraries and encourage to nurture a love of reading. Kids provide us with lots of excitement as they present drama, plays, and poems all focused on *International School Library Month Theme* of the particular year. Already, there are great plans underway in preparation to celebrate the *International School Library Month* this year, and support in any way is being sort for. 2016 looks like a very big event to our celebrations here in Zimbabwe. Below are a few photos of the past event in 2010: zimbabwereads.org/zimla/2011/10/01/school-library-month/

If they were to lay off the school librarians or to close down the school libraries completely in your part of the country, what kind of impact do you think it would have on the overall learning and recreational needs of the whole school community?

In one word it will be a DISASTER. The librarian and school library are the core components for creating and achieving a conducive learning environment. Even schools without a proper structure for a library have classroom libraries, thereby, exhibiting the need and importance of maintaining those two aspects: librarian and school library. The school library manned by a qualified school librarian provides essential research support and assistance in achieving higher learning grades. The library affords the students an opportunity to nurture a reading culture, explore the unknown; the possibilities are endless. Apart from supporting learning aspects the library is also a recreational space for such activities as storytelling, drawing and art, meeting new friends. The library supports the school curriculum by having additional learning and reading materials apart from those prescribed.

Regular classroom teachers versus and school librarians in your region, which one do you think would have a more optimistic and promising career path and career progression?

These are two distinct professions with totally different set of values. In essence, they complement each other and none is above the other in terms of recognition and importance. Both have different set of objectives to implement, manage, and achieve. The teacher, in my view, is there to implement the curricula, monitor and evaluate the student's progress with an aim to achieve high pass grades. As for the librarian's role, it is to support the achievement of learning objectives by providing a service that is conducive for further inquiry and recreational purposes.

Both professions have an optimistic, promising career path and career progression. They both provide avenues for personal growth intellectually and professionally. The opportunities are boundless which allows one to grow from entry-level to management activities. But, reality on the ground is that traditionally teachers have received more recognition than librarians. But with progressive school authorities, they have realized the enormous and invaluable contributions of librarians. Or, in the case in Zimbabwe, the training of librarians and teachers is similar being three to four years of certification to higher diploma. Further academic opportunities can be pursued at various institutions of learning to obtain grades as BEd, BSc in LIS, or MSc in LIS.

In addition, school librarians in general have a more promising career because they have more exposure to reading materials, there are more online courses one can embark on whilst at work, and local universities are introducing more part-time courses for continuous professional development of the LIS profession.

What is the situation for collaborations with other libraries or between school libraries in your part of the world?

Collaboration, whilst the most ideal way to go in order to share resources, is less practiced in Zimbabwean school libraries. However, it has not been practiced in my country, because there are no platforms for networking between school libraries like the case in developed countries, such as the UK. The most difficult part is that there are no strong linkages between school libraries that can enable them to come together, and to form a consortium to link and collaborate with school libraries in other countries. This will be a prolonged problem, as long as there is no School Library Board to coordinate activities in school libraries and help them collaborate.

Why is it so difficult to establish a school library association in your part of the country?

It is difficult because of the following reasons:

1. Very few schools have libraries.
2. The schools that do have libraries are not staffed with library professionals—some are even manned by groundskeepers!
3. School authorities do not value school libraries and librarians.
4. Reading and using libraries are not prioritized. Sports are actually more respected than a school library.

5. There is no government commitment devoted to school libraries.
6. The National Association for Librarians is not doing enough to establish an association for school libraries for our region.
7. School librarians struggle in terms of funding (working with tight budgets) and low salaries—without enough resources, they cannot run around. There is also an issue with the low number of trained professionals in school libraries, and only the top schools have the ability to attract and recruit qualified librarians.

It is, therefore, a critical time for the Ministry of Primary and Secondary Education to come up with a library service board to establish proper guidelines for school libraries in Zimbabwe. As long as there are no properly-run school libraries, it will be difficult to establish a professional association.

What is the situation of using the Internet and use of technology in the school library?

There are school libraries—particularly those in independent schools—that have come up with Media Resource Centers promoting the usage of Information and Communications Technology (ICT). In these schools, the libraries have gone all the way to purchase library management software like Oliver, Mandarin, and Koha to computerize their collection and come up with computerized cataloging of books and circulation management. However, government schools are not well-resourced, and they, by and large, are still using manual systems of loaning out books and recording them.

Hosea Tokwe
Chief Library Assistant,
Midlands State University1,
Library Department, Gweru,
Zimbabwe
Guiding an MA student
on how to write a research
proposal

Hosea Tokwe researching before preparing an information literacy session

IT IS A LONG, LONG WALK TO BOOKS AND VERY FAR AWAY FROM INFORMATION ACCESS AND DELIVERY: STORIES FROM SCHOOL LIBRARIANS IN ZIMBABWE

JERRY MATHEMA

School Librarian, Masiyephambili College,[1] Bulawayo, Zimbabwe, Africa

---

Please provide a brief self-introduction and tell us your professional and educational backgrounds. What did you study in university? Are you a second-career school librarian—meaning that did you have other careers before becoming a school librarian?

My name is Jerry Mathema. I am currently serving as the School Librarian of the Masiyephambili College—an independent coeducational secondary school that is based in Bulawayo, Zimbabwe, Southern Africa. I have a Postgraduate Diploma in Library and Information Science (LIS) from the National University of Science and Technology[2], a Bachelor of Arts in Media Studies from Zimbabwe Open University[3], a Higher National Diploma, a National Diploma and National Certificate in LIS from Bulawayo Polytechnic.[4] I am a second-career (school) librarian. I (previously) worked in another industry as a general worker before studying librarianship. I am the current Director of the International Association of School Librarianship for Region 1: Africa Sub Sahara, Chairperson of the

---

[1] Masiyephambili College—Homepage. Available at: http://www.masiyephambili.com/index.html.

[2] National University of Science and Technology—Homepage. Available at: http://www.nust.ac.zw.

[3] Zimbabwe Open University—Homepage. Available at: http://www.zou.ac.zw.

[4] Bulawayo Polytechnic—Homepage. Available at: http://bulawayopoly.ac.zw.

Zimbabwe Library Association[5] (Matabeleland Branch), and the External Assessor of Library, Archives, Records and Information Science at Higher Education Examination Council.

You previously worked in another (non-library-related) industry, could you tell me what kind of work you did before becoming a school librarian? Your previous working experience (skills and knowledge)—can it contribute to your current work as a school librarian?

I used to work as a spin cast operator: producing zinc buckles and other products for shoes, handbags, and belts. Yes, in a way it will inspire my students that you can do anything whilst you wait for your chosen career or encourage them to do something productive during one's gap years. It gives one some form of financial independence before one commits her/himself to a chosen career path.

Is that usual for a school librarian in Zimbabwe to have so many professional qualifications like you? With your qualifications, were there any reasons why you chose to work as a school librarian instead of becoming an academic librarian or working for the National Library of Zimbabwe?

Yes, it is usual to find most school librarians having so many qualifications especially if they first trained at the polytechnic. Most librarians will have a Higher National Diploma (HND)[6], National Diploma (ND)[7], and/or National Certificate (NC)[8] and then proceed to university to pursue a Bachelor of Science in LIS or in any other discipline that is related to librarianship such as records and archives, history, computer science, English and communication, media studies, education, and so forth. Let me hasten to say that the above levels are no longer of the same duration as was the case when we went to college. Higher National Diploma (HND) programs last for five years, ND programs for three years, and NC programs for one year. The Zimbabwean government has added new subjects, such as entrepreneurship, on-the-job training (one year), and national strategic studies

---

[5] Zimbabwe Library Association—Homepage. Available at: http://zimbabwereads.org/zimla/2012/07/09/zimbabwe-library-association-revives-branch-structures/.

[6] Higher National Diploma (HND) is a 4-year course of study in a given discipline like LIS.

[7] National Diploma (HD) and a 2-year course of study.

[8] National Certificate (NC) is a 1-year course of study.

(this course is meant to instill discipline and produce graduates that are patriotic to the ruling party).

During my training at polytechnic, I had the opportunity of working in all types of libraries (special, medical, academic, public, national, etc.). After graduation, I was employed by an informal institution that offered both secondary school and professional courses, and it lasted for nine months. I then moved to an independent trust school where I have worked for at last 11 years. I started off my career with an HND in LIS, and then I studied for a Bachelor of Arts in Media Studies and a Postgraduate Diploma in LIS.

There were no reasons why I chose to be a school librarian—I had no choice, and the only opportunities I got were in school libraries. Independent trust schools had lucrative salaries and perks at the time compared to academic and the national library. My biological children get subsidized education until they get to sixth form.

Choosing a career in school librarianship, was it an active choice out of personal interest? Or it was by chance and circumstance?

It was out of personal interest inspired by a former friend, who encouraged me to take up the course in LIS.

How did your friend encourage you to take up the course in LIS? What did he say to you that inspired you to choose a career in school librarianship?

He told me that librarianship is unique, demanding and also rewarding as a career. He told me that there were great opportunities of finding employment after training, and that the school librarian is the only personnel in a school setting who is acquainted to all the pupils and members of staff. The librarian promotes literacy, independent thinkers, researchers, and so forth.

All the things said by your friend—to encourage you to take up school librarianship as a career—did everything turn out to be true?

Yes, everything turned out to be true! I got my first job as a school librarian when I was doing my Higher National Diploma in LIS. I had to finish my diploma program through block-release study (or it is referred as part-time study). I am very popular with all the pupils at the school, and those who have already left still keep in contact with me. Some of them also donate generously to our school library. It is both exciting and rewarding to be a school librarian, as it affords me the opportunity to read constantly and widely.

The school librarians in Zimbabwe, are they mostly female? What is the gender ratio between male and female school librarians working in Zimbabwe?

I have not carried out a study regarding gender ratios, but it looks like most school librarians are women. Independent schools such as Christian Brothers' College[9], Masiyephambili College, Petra High School[10], Watershed College[11], Hillcrest College[12] are predominantly staffed by males.

Are you currently working as a solo librarian in the whole school?

Yes, I am.

Since you are working as a solo librarian in the whole school, how do you perform so many different tasks/duties at the same time?

Some tasks are routine and others are carried out as and when school opens, and some such as accessioning, cataloging, classification when library materials are acquired and added to the library collection, etc.

Could you describe the literacy rate in the whole country of Zimbabwe? Is education compulsory for all children (from what age to what age)? Would fines are imposed on parents or they would be imprisoned for not sending their children to school?

Zimbabwe has been a torchbearer when it comes to literacy since independence from Great Britain in 1980. For the last two decades, due to the economic meltdown, the literacy rate took a plunge, and we now have conflicting rates from different sources. Government sources put it on a high pedestal whilst anti-government sources put it at lower percentage. Yes, education is compulsory from the age of four (Early Childhood

---

[9] Christian Brothers' College—Homepage. Available at: http://www.cbcbyo.org/.

[10] Petra High School—Homepage. Available at: http://www.petraschools.com/.

[11] Watershed College—Homepage. Available at: http://www.watershed.ac.zw/.

[12] Hillcrest College—Homepage. Available at: http://www.hillcrestcollege.net/.

Development A). The constitution states clearly that children have the right to education, but fines are NOT imposed and there is no imprisonment whatsoever for not sending children to school. Human rights watchdogs and other organizations involved in childcare may prefer a charge against any parent who fails to toe the line if the case is brought to their attention.

Zimbabwe is leading the literacy rate in Africa, coming in at 91 % in the latest survey, despite a decade-long economic crisis that has impacted negatively on the quality of education.

The rankings published by the African Economist Magazine show Equatorial Guinea in second place at 87 % and South Africa in third position at 86 %. President Robert Mugabe is credited for Zimbabwe's high literacy rate after he declared education a basic human right following independence in 1980. According to the survey, low levels of literacy and education in general can impede the economic development of a country in the current rapidly changing, technology-driven world.[13]

Could you tell me the social and economic backgrounds from which your students come? In addition, what kinds of work do a majority of their parents do for living?

Upper-class, wealthy, and middle-class, Zimbabweans in the diaspora. The majority of the parents are business owners and executives, high-ranking government officials, university staff, other professionals who have educational benefits from their employers, and so forth.

Do your students have access to computers and Internet connectivity at home or they need to rely on the school libraries to provide them Internet services, computers, and basic reading materials?

Yes, about three-quarters have computers at home or use smartphones to access Internet services.

With reference to Zimbabwe's public school system, if the government's aims are just to eliminate illiteracy (teaching children how to read and

---

[13] *Zimbabwe leads literacy rate in Africa.* Available at: http://www.sabc.co.za/news/a/5c11890044581 c8fbd02fd744a7933f3/Zimbabwe-leads-literacy-rate-in-Africa-20140612).

write), and to prepare students to do well in public examinations—and if little is emphasized on inquiry-based learning—why do you think the local public schools should also be equipped with a proper school library that is managed by a qualified/professional school librarian?

- To compete favorably with independent trust schools;
- To have access to a wide array of both electronic and print resources;
- To gain literacy and media skills;
- To learn to do research and be independent information seekers;
- To improve numeracy;
- To acquire reading skills;
- To critically analyze information.

For the local public schools that have school libraries—please describe these public school libraries' resources (book collections, digital resources, library programs available, library furniture, and staffing situations, and so forth.) by comparing against the Masiyephambili College Library?

There is no library budget. They thrive through donations that are irrelevant to the school curriculum. The books are obsolete, there are no digital resources, and they are staffed by unqualified personnel or subject teachers who do not have qualifications in school librarianship. There are no library programs, the furniture is old and worn out. In some cases, the shelves are empty and the library has been turned into a marking room for English teachers.

Could you describe your typical day at work as a school librarian?

Collate statistics, housekeeping issues, shelving, shelf reading, charging and discharging library materials, accessioning, cataloging and classification, orientation and induction of new pupils and staff, supervising pupils, imparting information and media literacy skills, assisting pupils with the use of computers and helping them search relevant websites, library management and administration (budgeting, purchasing), and so forth. In addition, I need to answer reference questions face-to-face, through the telephone, using social media, draw up an inventory of new books, and other library materials, as well as to provide reader's advisory services, and so forth.

Do you need to take up any classroom teaching duties in addition to fulfilling your roles as a school librarian?

I am working as a full-time librarian, (and do not need to take up any teaching duties).

As a school librarian in your region, is there a nationwide or region-wide syllabus or curriculum that you need to follow, in terms of performing your work as a school librarian? If not, do you think it is feasible to implement a region-wide syllabus for school librarians? The absence of such a syllabus—do you think it is an advantage or disadvantage?

There is no nationwide syllabus or curriculum. It is feasible if we present our case to the relevant educational authorities. Through lobbying and advocacy, it can be possible.

What are the expectations amongst your students, other classroom teachers and the senior management in the school library, and in you—in the context of supporting the overall learning and teaching, as well as the development of other recreational activities of the whole school?

They expect the (school) library to support the whole school's curriculum, access to current information, as well as to provide unlimited access to a wide range of library services and materials—with the aim of supporting teaching and learning of the school community as a whole.

Please give a list of successful library programs (supporting students' overall learning and teaching of other teaching staff) initiated by you as a school librarian?

- Information and media literacy skills,
- Writer's and reading club,
- International School Library Month celebrations,
- School outreach program with staff and pupils.

What are the major challenges and difficulties faced by you as a school librarian?

Budget constraints, little (physical) space for studying and reading inside the school library, and no study center for the sixth-form students.

Which parts of your job as a school librarian do you find most rewarding?

Automating school libraries and integrating ICT (Information and Communications Technology) with other library services. Teaching library and information skills to the students.

The professional knowledge, skills, roles, and other job-related competencies for a school librarian—have they undergone major changes in your region in the last five to ten years? In your opinion, what is the future for school librarians in your region?

YES! They have changed tremendously; the school librarian is now integrating ICT with other library services to increase patron access to a wide array of the information sources that are available. I believe that the future for school librarians is very bright in the region as long as the economic situation improves. The polytechnics and universities that teach library and information science have since changed their curriculum to align with emerging technological trends within the field of school librarianship.

As a school librarian in Zimbabwe, do you sometimes feel that you could choose to work very hard or do nothing at all—at the end, you would still get paid the same amount of salary? People are sometimes promoted because of their seniority (only they have been here longer), and not because of how well they do their jobs?

I choose to work really hard. There are instances where corrupt heads promote someone not because of their seniority but because they are friends, relatives, lovers, cronies, and so forth.

Throughout your career as a school librarian, did you ever have any regrets or second thoughts?

No!

If they were to lay off the school librarian or to close down the school library completely, what kind of impact do you think it would have on the overall learning and recreational activities of the whole school community?

The pupils will be information-starved. They will not have access to free Internet services, expensive books the library has in its stock (reference material), a conducive, comfortable reading and study center, etc. The pass rate will plummet to low levels. Literacy rate will also be affected.

School libraries/school librarians and inquiry-based learning—do you think they go always hand in hand? In the school environment, true inquiry-based learning could not be carried out without a proper school library that is managed by a professionally trained school librarian?

Yes, they do go hand in hand. As a trained school librarian one has the skills to carry out a successful and satisfactory inquiry-based learning program.

Regular classroom teacher versus and school librarian in Zimbabwe, which one do you think would have a more optimistic and promising career path and career progression?

School librarian. A school librarian with a diploma stands a chance of progressing to higher learning academic institution than a teacher who will need master's degree to be part of a university lecturing staff.

Why do you need a study center dedicated for the sixth-form students?

Sixth-formers are the most mature students at a school and they need to meet, discuss, and debate on serious academic issues without any disturbances from lower forms. Through inquiry-based learning they may have immediate responses and more time to interact with the school librarian. It inculcates a culture of independent research and resource sharing.

Jerry Mathema
School Library Media Specialist at
Masiyephambili College Library &
Media Center, Bulawayo, Zimbabwe,
Africa

Cambridge advanced-level students studying at Masiyephambili College Library &
Media Center

Audio-Visual Section of the Masiyephambili College Library & Media Center

CHAPTER 4

MY LIBRARY, MY LIFELINE

DUMEBI EZAR EHIGIATOR

The Vale College,[1] *Ibadan, Nigeria*

Please provide a brief self-introduction and tell us about your professional and educational backgrounds. What did you study at university? Are you a second-career school librarian—meaning that did you have other careers before becoming a school librarian?

My name is Dumebi Ezar Ehigiator. I earned my first (Bachelor's) degree in Library, Archival and Information Studies, a Master's degree in Library and Information Science (MLIS), and currently undertaking a doctoral degree program in School Media at the Centre for Educational Media Resource studies, University of Ibadan.[2]

From which university did you earn your first (Bachelor's) degree in Library, Archival and Information Studies, a Master's degree in Library and Information Science (MLIS)?

From Imo State University[3] and University of Ibadan, respectively.

What are the minimal professional qualification requirements for working as a school librarian in Nigeria?

Candidates must possess High National Diploma (HND), Ordinary National Diploma (OND), National Diploma (ND), or Diploma Certificate in Library Science from a reputable institution. In addition, candidates must have five O level credits, including English Language at not more than two sittings.

Are all local schools in Nigeria also equipped with a school library that is managed by a qualified school librarian?

No!

Choosing a career in school librarianship, was it an active choice out of personal interest? Or it was by chance and circumstance?

I have always desired to be a librarian because I had an amazing librarian in secondary school. She instilled the love of books and the library in me.

How did this secondary school librarian instill the love of books and the library in you? Was that her professional working attitude or something special that she said to you personally?

My secondary school librarian, Mrs. Edet, is the most amazing person to know. I will always remember the first time I went to the library. She set up my first library card and walked me around the fiction section to find the perfect first book. *The Famous Five*[4] became one of my favorites and I renewed the series many times as a child. Throughout the years in school, she made the library like a second home to me. She made me write book summaries, and told me I would be a brilliant writer. Today, I am.

Mrs. Edet—could you tell me about her cultural and educational background? Could you also describe her teaching style and what make her so different from other librarians and teachers whom you encountered in the past?

She was a black American married to a Nigerian. She had a Master's degree in Library Science. She understood my learning style, and she was focused on helping me achieve all I could. I also remember countless times that she pushed me…to think more deeply, to write more critically, to express myself more fluently. She encouraged me to believe in my own abilities but never let me make excuses for failure, for simply not trying, or for being lazy in my mind and work ethic. It is not that she was "the nicest"

---

[4] *The Famous Five* is the name of a series of children's adventure novels written by English author Enid Blyton.

or "the friendliest," but I had nice and friendly friends and family. I needed a strong mentor who would not tolerate anything less than my best effort.

Most people would define a good teacher as someone who makes their students excel academically and do well on their tests. I believe that is almost right, but it is a little off. I believe that a good teacher does not have one dimension but two. They not only make you excel, but they also make you want to go to school. They care about students' insecurities and problems, and, most importantly, they are there to support you. She was all that.

What is the official language of the country of Nigeria?

English is the official language here.

What is the average literacy rate amongst population in your region of Nigeria?

I cannot give an accurate figure. But, it should be hovering around 50%.

In your region of Nigeria, would a majority of your students end up going to the university after finishing high school?

Yes, but maybe not immediately.

What are the social backgrounds of your students? What do a majority of their parents do for living?

I work in an elite school. Parents are high earners.

Could you describe the Internet infrastructure in Nigeria? If the Internet network is very much developed, would it be more effective to bring iPads or digital tablets to the students—thereby enabling them to have easy access to large amount of reading materials via the Internet, instead of investing a great deal of manpower and resources to build a physical library on school campus?

Nigeria's Internet sector has been hindered by the country's underde-veloped and unreliable fixed-line infrastructure, but this is changing as competition intensifies and new technologies are able to deliver wireless broadband access. Phones and tabs are not allowed in my school because of the distractions therein, but students have access to the computers and electronic databases in the school.

Are you currently working as a solo librarian in the whole school?

Yes, I am.

Could you describe your typical day at work as a school librarian?

My typical day cannot be distinctly described. I do a whole range of activities each day. I do things like ordering the books that get to the shelves, weed out the old books to make room for new ones, make sure that the books are in the right section of the library, work on complicated reference questions for students and teachers, teach new technologies, arrange programming and professional development sessions for the teachers, maintain the school's blog, attend classes, and perform customer service all day long.

Do you need to take up any classroom teaching duties, in addition to fulfilling your roles as a school librarian?

Yes, I do.

As a school librarian in your region, is there a nationwide or region-wide syllabus or curriculum that you need to follow, in terms of performing your work as a school librarian? If not, do you think it is feasible to implement a region-wide syllabus for school librarians? The absence of such a syllabus—do you think it is an advantage or disadvantage?

In Nigeria, there is none. But, the Centre for Educational Media Resource Studies in Oyo State is working with schools to implement one. The absence is a disadvantage. The library curriculum will provide a varied and extensive range of activities related to literacy development and reading promotion for the students.

What are the expectations amongst your students, other classroom teachers, and the senior management in the school library, and in you—in the context of supporting the overall learning and teaching, as well as the development of other recreational activities of the whole school?

The role and perception of my library is focused on the expectation that I am a learning enabler. I am to model and encourage all staff as readers, model and encourage digital citizenship and create a safe and supportive environment for my students to explore new ideas and embrace change.

Please give a list of successful library programs (supporting students' overall learning and teaching of other teaching staff) initiated by you as a school librarian?

1. Reading schemes for students and parents,
2. Book exhibitions,
3. Seminars and workshops on reading conducted by teachers or local writers,
4. Reading-related competitions,
5. Book Club for students: Meet during lunch to read any type of book and talk about it,
6. Book Mobile: Deliver books in the summer to areas that do not have access to books,
7. Movie Nights: Show movies in the library,
8. Reading Extravaganza (World Book Day) Celebrate reading with Games, Crafts, Food and Drinks, Drama and Dance.

What are the major challenges and difficulties faced by you as a school librarian?

It is quite taxing to find better and more effective ways of engaging school with school board members, teachers, students, and parents in honest conversations about librarians as instructional partners. How do I engage them with the shared story of the library I am trying to compose and construct with our teachers and students? In my heart, I still believe in the possibilities of libraries and school librarians—but, they will never come to fruition if we acquiesce and abandon the effort to elevate the library as a site of participatory culture and a cornerstone of every child's learning experience in schools, as a partner who can support our teachers by being embedded as part of the team to give every child positive, constructive, meaningful learning experiences, and so forth.

The inescapable conclusion is that when one considers all the evidence of advancing technology, education reforms, societal changes, information literate customers, and globalization of "everything" and their impact on librarianship and libraries, it is crystal clear that 21st century librarianship must be drastically different from all previous concepts of librarianship. It requires a professional who embraces the potential of technology, creatively finds appropriate ways to implement it into library services, and one who has more diverse—even "unconventional"—skills than ever before. The 21st century librarian is a professional who understands the millennial library customer, is able to adapt existing services and create new ones to meet their community's needs, and change the public perception of "library." Management is tricky because sometimes I feel like I have to fight for absolutely everything I want.

Why do you think, "Management is tricky because sometimes I feel like I have to fight for absolutely everything I want"? Is it because of a shortage of money or because the school's senior management does not think that the school library is important amongst the whole school community?

Books do not come cheap here. And, as much as I would like to have new books every session, it does not always happen. There are other areas in school that also require more funds. Most times, I have to make more and more cuts in my own budget to accommodate others.

However, I am fortunate in having the support of senior management and holding a subject leader position, which helps give me more clout. But, many school librarians are seen purely as minders of a spare IT suite or as date label stampers. They are enormously, depressingly, frustratingly underused.

Do school librarians in Nigeria get paid the same amount of salary as other subject/classroom teachers?

I cannot speak for others. But, I know I am paid more than most subject and class teachers in my school.

Which parts of your job as a school librarian did you find most rewarding?

My predecessor was not a very friendly fellow. In fact, I heard she hated people coming into the library. I love it when teachers come up to me, and tell me how nice it is to be allowed to be in the library, and how helpful my staff and I are. Also, when parents of students come in and give me compliments about how much their children enjoy library time. I love anything that encourages and fosters a love of reading.

You said your predecessor was not a very friendly fellow and she hated people coming into the library—why was it the case? Was it because more people going into the library would simply create more work for her?

She did not want to be responsible for books missing. So, it was safer to have an arranged library with no user. Safer and cheaper, I guess.

The professional knowledge, skills, roles, and other job-related competencies for a school librarian—have they undergone major changes in your region in the last five to ten years? In your opinion, what is the future for school librarians in your region?

Of course, they have. In the past, librarians curated information from foreign creators and disseminated it to a local community. Modern librarians curate local information and disseminate it to a foreign community. The flow of information has flipped. The altered role of libraries is a great opportunity to showcase African knowledge. Getting information into the world is easier and cheaper than ever. African libraries need to take up the responsibility of being partners in information creation. This means that policies must be altered—and, of course, that budgets must be increased. University leaders, decision makers, governments, and library users need to understand and support the changes that are reshaping libraries. Librarians, too, must embrace these changes. They will need new skills to support the creation of information. Many library schools are already responding to these new needs by offering advanced degrees in digital curation. It will also be important to reconsider the very physical space of a library. Paper and glue book collections are shrinking and, in some libraries, disappearing.

If they were to lay off the school librarian or to close down the school library completely, what kind of impact do you think it would have on the overall learning and recreational needs of the whole school community?

Students' achievement level would drop considerably. Students are more likely to succeed when they have libraries that are well-staffed, well-funded, well-stocked, technologically well-equipped, and more accessible. And the neediest learners benefit from trained librarians and quality library programs.

As a school librarian, you could choose to work very hard or do nothing at all—at the end, you would still get paid the same amount of salary—do you agree?

I would agree…if I did not love my job.

School libraries/school librarians and inquiry-based learning, do you think they go always hand in hand? In the school environment, true inquiry-based learning could not be carried out without a proper school library that is managed by a professionally trained school librarian?

When it comes to effective teaching inquiry-based learning, an educator's best bet is to pair up with their school librarians to snag students' attention early, as it important to develop essential questions that connect the standard to the real world. Connecting learning to the experience of the learner makes it more relevant and allows students to manipulate and apply their learning in ways that they can see. This approach focuses students' attention and immediately distinguishes the learning from a simple bureaucratic task that they just have to get through.

Having a passion for school library work—do you think it is something that is inborn (some people would say it a calling) or it is something that could be developed over experience and exposure?

Both.

If a young person in Nigeria who is inspired to choose school librarian-ship as his/her lifelong career—what would you say to him/her?

Advice to aspiring librarians: The patron (customer) is not always right. Many business ideas are applicable to libraries, but this one bugs me. Be clear, concise, courteous, and reasoned in disagreements. However, bad behavior from patrons should not be rewarded.

Second, be flexible and adaptable: change is a certainty. While antici-pating change has its value, sometimes the best thing we can do is to be open and ready to roll with it. This is not easy, especially at first. Yet recognizing this and giving yourself a chance to process the discomfort with an objective eye can help. Adopt a wait-and-see attitude about where it might lead before determining that it is bad. If possible, look for places where you can take some control and affect your own change.

Our professional growth is organic by nature and often takes us in a direction we did not expect. Since beginning my MLIS program two years ago, I have had the opportunity to talk to a number of people who are, themselves, considering getting an MLIS (or the equivalent). I am always happy to talk to them and give them advice and information, but I always cringe when any of them tell me they want to be a librarian because they love reading (or some variation), because I think it is such a bad reason for becoming a librarian.

However, I think that loving to read is a good quality in a librarian, but it does not seem like it should be the main reason one chooses the profes-sion. (I do not know of any librarians who actually get to sit around and read all day, and the ones who do spend a fair amount of time reading do not get to choose the material).

So, I am trying to come up with motivations that I think more realisti-cally fit the typical work a librarian. For example, "I want to be a librarian because I love storytelling and working with kids." Or "I want to be a librarian because I love organizing information and figuring out how to present it in an intuitive way." Or "I want to be a librarian because I like working with people to help them find the resources they need."

FURTHER READING

A Conversation with Dumebi Ezar Ehigiator, Author of "The Spider's Web." http://pulse. ng/prose_poetry/debut-nigerian-author-a-conversation-with-dumebi-ezar-ehigiator- author-of-the-spiders-web-id4426543.html.

Dumebi Ezar Ehigiator
School Media Specialist at
The Vale College, Ibadan, Nigeria

The Vale College, Ibadan (Oyo State, Nigeria)

The Vale College students working on a group project in the library

PART II

Asian

CHAPTER 5

CHILDREN EXPERIENCING
THE JOY OF READING IN JAPAN

MAMI KOBAYASHI

Elementary and Junior High Schools, Fukushima, Japan

Please provide a self-introduction, including your professional and educational backgrounds. Were you a second-career librarian?

My name is Mami Kobayashi and I am the school librarian in the town of Yabuki, Fukushima Prefecture.[1] This is my 6th year of working as a school librarian. I earned a degree in Art History in the School of Humanities at Koriyama Women's University.[2] In the School of Humanities, you could get certification to work in libraries, become the director of social education programs at town public halls, or become a museum curator.

Before I became a librarian, I had worked a few different jobs completely unrelated to librarianship. The jobs I held were mostly in the customer service industry like restaurants and cosmetic stores.

How did you come to work in your current position?

I actually got my start in the field with thanks to a friend. I remember telling them that because I had always wanted to work in a library, I got my certification. Three years later, that friend told me that the current town I am working in was looking for a qualified librarian to work in the town's schools and asked if I was interested. I cannot tell you how excited I was that my wish was finally coming true. It just goes to show you that if you plant the seeds, something will grow and come off your hard work.

---

[1] Yabuki, Fukushima Prefecture–Available at https://en.wikipedia.org/wiki/Yabuki,_Fukushima

[2] Koriyama Women's University–Homepage. Available at: http://www.koriyama-kgc.ac.jp/

In Japan, how can one become a school librarian?

In Japan, there is actually no certification for specifically a school librarian. Once you are certified, you simply are a librarian. There are qualified for several ways to become a librarian. Certification can be completed at a university or two-year junior college. There are also correspondence courses available for those who are working or want to study from home. For those who have already graduated from university, every summer at designated universities, lecture-based courses are offered for librarian certification.

Are you currently working as the only school librarian?

Currently, there are two librarians working in my town, including myself. We share one junior high and four elementary schools within the town. I usually visit each school twice a week.

Please describe a typical day for you as a school librarian.

I work from 9:00 a.m. until 4:00 in the afternoon. Each day's work usually consists of arranging the bookshelves and making sure everything is organized. I also check books in and out when the students come into the library and update the data on what books are currently checked out. I also make and decorate book displays. Depending on the season, I like to change the theme and pick featured books accordingly.

Whenever students come in during recess or break times, I try to be a friendly face to talk to and recommend books to them. During the periods that whole classes are using the library, I prepare whatever materials the teacher needs and act as a support. When I have time by myself in the library, I look at publishing websites to find what new books are out there that the students might be interested in.

What are the expectations amongst your students, other classroom teachers and the senior management in the school library, and in you—in the context of supporting the overall learning and teaching, as well as the development of other recreational activities of the whole school?

While there are not requirements other than the ones I listed previously, when I am in the library I make sure that students who are being noisy quiet down. I also clean with the students during cleaning time after lunch. I meet with teachers and help plan lessons that take place in the library. I also meet with the teacher who is in charge of the library in each school as well as student-run library councils to discuss any library-related events. I do my best to help the teachers make use of the library and the collection for their classes. In order for classes in the library to run smoothly, it is important to plan together and prepare the necessary materials. However, because the days I go to each school are limited, it can be a little difficult to allocate time to sit down and plan a lesson together.

In Japan, is it legally required that there must be a school librarian in the schools?

In Japan, there is the national School Library Act, which according to Article 6, originally stated that there should be a school staff member with librarian certification to manage the library and facilitate library activities for the students (currently, only 50 % of the local schools in Japan have a school librarian). Until 2015, it did not specifically state that, the person must be a school librarian. There was a huge movement for change, however, and within the revised school act the word "school librarian" was used for the first time. Local municipalities began accepting this change and actively worked towards employing school librarians. In Fukushima prefecture, schools are aiming to have all school libraries managed and run by qualified librarians. Article 6 also states that there must be either national or local efforts to enhance the quality of school librarians.[3]

In order to fulfill that part of the School Library Act, the Board of Education in the town I work for meets with the school librarians and the teachers at all of the schools who help manage the library when the school librarian is not there. Together as a group, we discuss how to best use the school libraries best and address any issues at the individual schools. In order for the libraries at each of the schools to be actively used, I think making time to sit down and share ideas is really so important.

---

[3] Japanese School Library Act. (Japan April 1, 2015)
http://law.e-gov.go.jp/htmldata/S28/S28HO185.html

Does your town use a mobile library? What do you think are the benefits of mobile libraries even when there are libraries in the school?

My town employs the use of a mobile library. The mobile library serves the town's elementary schools, kindergartens, and nursery schools. Since it is different from a public library, community members cannot take out books from the mobile library. The school librarian is expected to check books in and out at the mobile library when it visits the schools and we make book recommendations to the students if they ask. We try to bring books that are not available in the school libraries and students always get so excited for the mobile library day.

For elementary school students, how far they can travel is quite limited. So students who live far away from the public library are just not able to get there by themselves. By taking the library to the students, we give them even more opportunities to develop their love for reading. I think providing that for the students is so important.

Do you work closely with the town library?

Whenever a teacher wants to use a book in school but it is not available in the school library, I work with the town library to get that book for them. I think that cooperating with the town library is definitely important.

Are there any library organizations that you are a member of and if so, does that membership benefit your work?

No, I am not a member of any official organizations. However, as I stated previously, I take part in the town's meetings with the board of education and teachers to discuss our school libraries.

What are some successful library programs initiated by you or your colleagues in the classroom or the library?

One of the things I changed was to increase the chances for students to borrow books. Originally, the times for borrowing and returning books was limited to only when the students in the library club were managing the front desk during recess. There were a lot of students who ran out of time to both play with their friends and borrow books. I wanted them to be able to use the library even more, so I suggested that students can borrow books anytime the librarian is in school.

A new project I started this year was to increase reading at the junior high and get as many students as possible in the library. I got inspiration from point cards that are given out at stores and decided to give every student their own library card with space on the back for stickers. The more books the students read, the more stickers they got. They loved the challenge of collecting as many stickers as they could. The students could really see their own reading progress by counting the number of stickers they received. It was also fun to see them showing their cards to their friends and competing to see who could read more books and get more stickers.

What are the major challenges and difficulties faced by you as a school librarian?

As a librarian, my job is not only to check books in and out but to be a support for the teachers when they use the library. However, because I do not always know the ins and outs of every subject for every grade, it is necessary to meet with the teachers and request lesson details so that I know what and how to prepare. While I have not experienced major difficulties, one challenge that I have faced is how to be the best support for the teachers. Since I am always in the library, sometimes I am not sure just how to manage certain students. Whenever that happens, I consult the teachers for the best way to support them when they are leading the lesson.

Which parts of your job as a school librarian did you find most rewarding?

One of the joys of this job is after I recommend a book to a student they come back to tell me how interesting it was. When I share books with students, I feel like I am able to build a closer librarian-student bond with them. I am also delighted when a lot of students visit the library. Seeing the students come into the library, immerse themselves in a world of books, and enjoy the library space in their own way is one of the most heart-warming parts of this job.

Through this job, one of the things I have learned is just how much I want children to experience the joys of reading. Nowadays, so many children are constantly online or playing games. But through reading, they can develop a strong imagination and learn so much. Through my work, I have to share with my students the importance of reading. A library is not just a place where books are: it is a place where children can feel they belong. The library is where I am able to really reach out to the students with special needs and students who are having a tough time.

What kind of attributes does a motivated and successful school librarian always possess?

I think while it is obviously important for school librarians to have a love for books and enthusiasm for the job, they are not the only important attributes. I think the ability to communicate with both students and teachers and to have a love of people is so important. It is also important that librarians are really in tune with the needs of their patrons. While I visit multiple schools, I think for me it is important to really think of myself as a member of each school and to do my best for each student and teacher.

What do you think about the future of school librarians in Japan?

I think the future of school librarians is quite positive. The reason being, as I touched upon earlier in the interview, is that thanks to changes in the law, the necessity and importance of school librarians was recognized. That being said, because the availability of jobs completely depends on individual municipalities, there is the issue of just how much room for change there really is. However, there is a movement to establish school librarianship as a specific certification. I think it will be important to see how things regarding school librarians in Japan continue to develop.

If the school decided to lay off the school librarian or to close down the school library completely, what kind of impact do you think it would have on the overall learning and recreational activities of the whole school community?

Since teachers who manage a homeroom in addition to their regular workload are so busy, is it really hard to find time for them to manage the library. That is why it is so important to have a librarian. If there were no librarian, the library would simply be a locked-up room in the school or simply a place to store books. If there were no library, there would be lesser chances for students to enjoy reading and there would be the loss of space that students need to study. Most importantly, without the library they would lose a key space for them to not only nourish their minds but also their hearts.

Do you have any interesting stories from your time as a librarian that you would like to share with the readers?

In my fourth year as a librarian, I started storytelling sessions at the library in junior high. I gradually started to see students come in on their own and grab a book and read together. One male student, who had the tendency to act out at times, would also come in. At first, he would take a book and half-jokingly read aloud to his friends. His friends would also joke around while listening. But, I noticed that the more these boys read together, the more they became serious as both storytellers and listeners. This actually prompted other students to gather around and listen in. When I saw this, I was so moved by the power that books and picture books have.

CHAPTER 6

BRINGING THE CONCEPTS OF U.S. PUBLIC LIBRARIANSHIP TO A SCHOOL LIBRARY IN HONG KONG

JUN NIU, TAMMY NG, JOLI MOORE

School Librarians, Po Leung Kuk Choi Kai Yau School, Hong Kong, China

The following interview was originally published in *The End of Wisdom? The Future of Libraries in a Digital Age.* Amsterdam: Elsevier. (2016). Reprinted with permission.

Could you take turns to introduce yourself, particularly your education background and your professional training as school librarians?

Tammy Ng (TN): I am Tammy Ng. I am one of the teacher-librarians at the Po Leung Kuk Choi Kai Yau School (CKY) (保良局蔡繼有學校).[1] My core duty is to take care of the primary school section, that is, students from 1 to 5 years. I graduated in United Kingdom (UK) in 2004, and my undergraduate major was English Language Teaching. I graduated with a Master's degree in Library and Information Science (MLIS) from The University of Hong Kong (HKU)[2] in 2010. On top of that, I also earned a teaching certificate in Hong Kong. Before coming to CKY, I was working as a teacher-librarian at a Primary Years Programme (PYP) School. This is my fourth year in CKY, and 2014 would be my tenth year working as a teacher-librarian in Hong Kong.

Joli Moore (JM): I am Joli Moore. I mainly deal with 6 to 10 years students. I am from Hungary and got my Bachelor's (degree) in Hungary, majored in Literature and Linguistics, and minored in Communications. I got my Master's in Library and Information Science in the USA. Before I started

---

[1] Po Leung Kuk Choi Kai Yau School—Homepage. Available at: http://cky.edu.hk/
[2] The University of Hong Kong—Homepage. Available at: http://www.hku.hk/.

working at CKY, I was employed at the American International School (in Hong Kong), where I worked as an Elementary School Librarian. In fact, this is already my fourth year working as a teacher-librarian in Hong Kong.

Niu Jun (NJ): My name is Niu Jun, and originally I am from Mainland China. Before coming to Hong Kong, I lived in the United States for many years, and I got my MLIS degree from San Jose State University. Before moving to Hong Kong, I worked for several public libraries in the United States. CKY is the second school I have worked for in Hong Kong. The first school was the Independent Schools Foundation (ISF) Academy.[3] Last year, I finished my second Master's degree in Chinese language and literature. It was really helpful in terms of assisting my colleagues (especially Chinese teachers) with their subject teaching. My library study focused mostly on English-language literature, as it was very helpful on collection development for our jobs at the public libraries.

In Japan, also in China and in Hong Kong, many schools cannot even afford to have one full-time librarian on staff, but this School (CKY) can have three full-time professional librarians working concurrently side by side, under one single school library—could you tell me about the situation at CKY? What objectives and results are you trying to achieve for having three librarians working together within the same school?

NJ: I think it depends mostly on how you expect the library to function within the school community as a whole. If the library only serves as a storage or warehouse for printed books, and performs merely the check-in and check-out functions, you do not really need a fully trained professional librarian with a MLIS degree; a high-school or college graduate with years of experience would be more than sufficient. However, when it comes to implementing information literacy (IL) or reading advisory—they would definitely require someone with much more training and special professional skills to deliver these lessons. For the other local schools in Hong Kong, they might be perfectly operational with just one single teacher/ librarian overseeing all the daily operations of the entire school library, but they would not be able to achieve all the things that we are currently doing at CKY.

---

[3] The ISF Academy—Homepage. Available at: http://www.isf.edu.hk/en.

At the other local schools in Hong Kong, it is almost mandatory that prac-ticing school librarians are also certified teachers. So what is the situation here at this school? To work as a school librarian at CKY, is having the MLIS degree more important than the teaching diploma?

TN: In my personal opinion, both the teaching diploma and the MLIS qualification are equally important, especially when you are working for a school, in which we (school librarians) need to conduct a variety of outreach and educational activities for the students. For example, in addition to overseeing the daily operations of the school library, you also need to have the basic knowledge and skills to deal with issues such as psychological problems and disciplinary issues amongst the young students.

More importantly, you also have to understand the whole curriculum, the needs of the other classroom teachers and what they are doing as well, and so forth. For these reasons, as a teacher-librarian you need to have a much wider overview of the whole school community—because we librarians have a much bigger, and in fact a very influential role to play. Supporting the information and reading needs of the students is just a very small part of our many duties. Hence, we librarians have to know every-thing that is going on within the whole school.

But in reality, it is kind of difficult in Hong Kong, because there are not many universities that offer MLIS programs. In the context of Hong Kong, it is sometimes unrealistic to expect practicing librarians to have both the teaching diploma on top of the MLIS degree.

Since the three of you come from very distinctive cultural and educa-tional backgrounds, how do you complement each other? How do your different experiences contribute to the school library programs as a whole? For example, NJ previously worked for a number of public libraries in the United States before coming to Hong Kong, while TM and JM have more training as classroom teachers and school librar-ians—so how do you complement each other? Maybe you could tell me more about the organizational/staffing structure of this school library at CKY?

TN: That is why our job as teacher-librarians here at CKY is so interesting. In fact, during the initial stage, it took us a lot of time to get to learn about each other's strengths—that is why we have such clear-cut job descriptions amongst the three of us.

Tammy is in charge of primary division, Joli is responsible for English teaching from 6 to 10 Years in the secondary division, and Jun supports the International Baccalaureate (IB) Diploma Program (DP) and Chinese teaching in the secondary division.

Based on my observation, the students at CKY seem to be coming from multicultural from multicultural backgrounds—do you think the information needs and reading interests amongst students of different backgrounds also tend to differ? Could you tell me about your experiences here at CKY?

TN: Dealing with students from multicultural backgrounds could be a big issue when it comes to collection development. For example, when we are placing orders for new books—we sometimes encounter titles that touch on issues regarding sexual orientation of young people, for example, issues such as homosexuality could be controversial amongst religious families or Chinese families with very traditional values…as a result, we would have to discuss with each other before any purchasing decisions are made.

Do you mean students and parents from different cultural and ethnic backgrounds perceive issues relating to sexuality very differently? Do you mean one of your duties as a teacher-librarian here is to use your professional judgments to exercise censorship of any new titles coming into the library?

NJ: Well, I would say it has more to do with how to select the most suitable books for your targeted group. In other words, we make an effort to select books that are appropriate to our students' needs in the local "Hong Kong context," rather than using the word "censorship."

No one can deny the fact that homosexuality does exist in our society, but we also need to know that we are in Hong Kong; and a majority of our students are from local Chinese families. This is not to say that none of the books in our school library touches on those topics—but we obviously need to be careful how students are educated in support of their sexual education along with the guidance and counseling department.

At CKY, both the primary and the secondary students are sharing the same school library—and one of the unique features of this school is that students at all ages and levels do love reading, and our school does put a lot of emphasis on the teaching of both the English and Chinese languages. We want our students to be fully bilingual.

Can you give me examples of successful library programs launched by your team of teacher-librarians?

NJ: We do quite a lot actually. We really try to make our school library busier and busier every day. Yesterday, we sat down together and counted the total number of (library) programs that we accomplished in the past year which were reading competition, book swap, book illustration, student librarians, parent volunteers, Christmas card design, and you name it. Since last year, we have invited several Chinese and English authors to come to give talks to both out primary and secondary students.

In terms of your library programs, do you focus mostly on author talks?

NJ: No, it is just one of our many programs launched by us librarians. We have another program called *Battle of the Books* catered for the elementary students—it is in fact a very popular reading incentive program amongst the local international schools in Hong Kong. Even though this was the very first year for our school to be taking part in this program, we made it into the final and volunteered to be the host of the semifinal competition (of the *Battle of the Books*). For this semifinal, quite a number of schools came to our school to compete with our students.

Another activity we have at our School is called the *Book Swap*—that is, after the children finished the books—provided they are still in good conditions—they are encouraged to swap with their friends. And we also have student librarians (program). Even though we have full-time library assistants working at our library, we still want to bring up the "the leader-ship role" amongst the students, that is, we want them to take part in some-thing that could contribute to the school community as a whole, and also to learn something (for example, sense of responsibility and group work, etc.) via performing these library duties. In fact, they did rather a good job in terms of keeping our bookshelves tidy as well as setting up the posters.

School librarians are not considered subject leaders in the traditional sense, as they do not need to prepare students for any major or public examinations. Additionally, compared with other PE or music teachers, their contributions achievements could be easily reflected in the numbers of awards or prizes won by their students. On the other hand, it does not seem to be the case for most school librarians. And a large number of teacher-librarians are suffering from a lack of recognitions, a lack of support from the classroom teachers, as well as from the senior management. In your opinion, do you think such unfortunate situations are caused by the fact that the quality of work and success of the school librarians are difficult to measure, especially in quantifiable terms?

NJ: If your successes are measured strictly by high school grades, students' prizes, awards, or medals, and so forth—yes, they are indeed hard to measure in terms of library input. But for the *Battle of the Books* event that took place in May (2014), I think we would definitely call it a "success"! But such difficulty in performance measures is not only found in school libraries. When I was working for a public library back in the United States, it was equally hard to measure the success in quantitative terms either. But I remember this public library manager in US telling me, *"Last year we were the third busiest (public venue) in town—the first one was COSTCO;"* and the second one I forgot. She said in the following year, *"I want to be number two!"* You know COSTCO is one of the most popular and busiest supermarkets in the city, and a lot of (American) people like to go shopping there. At that public library (in US) which I worked for, they would use the automatic door-count system to calculate the total number of entering the library each year. So what I am trying to say is that as long as the students still enjoy reading; still come to the school library and continue to use our services and resources voluntarily, we are already "successful"! Most importantly, it is our main role and core responsibility to foster students' motivation towards voluntary reading, as well as to stimulate their interests for self-learning—especially self-learning that would take place beyond their routine classroom environments.

Under the current digital era, there is so much reading material readily available on the Internet. Given that there is already so much of such material that could be downloaded for free from Google—what roles do the school libraries and school librarians play nowadays—in terms of contributing to these young students' overall learning and reading experiences? Does your school library still place heavy emphasis on the printed resources?

NJ: Hong Kong is a small place, but there is already a variety of schools, communities, and families from different social, cultural, as well as ethnic backgrounds. So how could we accommodate all their different needs and interests? Yes, we still have a huge collection of printed books. But on top of that, we also have a rich collection of e-books and other online audiobooks for the students, as well as their parents to enjoy.

Currently, we have over 250 (English language) titles of online audiobooks, and the downloading rates are really fast. We bought these online audiobooks from a US publisher, and they were just made available online for our students not long ago. Both students and parents could use their smartphones or iPads to listen to them via online streaming. The reason for selecting these online audiobook titles was because a majority of our students are from local Chinese families, and their parents would like them to acquire better English pronunciation, as well as listening comprehension skills.

Soon after I made these audiobook titles available online (for the students), I was surprised there were already 900 downloads within the very first week. Since we only have about 900 students in total in our primary school, such high downloading statistics already meant that both the students and their parents were already using these audiobooks actively; and we (school librarians) have done something "right" for the students. So I think this is a really "positive sign" that the parents also liked what we have provided for their children, and they have been also supporting what we are doing in the backstage. Indeed, we librarians have spent a lot of efforts on promoting these newly acquired audiobooks—by putting up signs everywhere inside the library—constantly reminding the parents about the availability of these online audiobooks.

Do you have any good suggestions or strategies in terms of requesting the senior management to grant you extra manpower or resources for implementing any new programs or services for your library?

NJ: I think the most important thing for us (librarians) is to find out, as well as to understand the missions of the school. In our case, we first have to understand that it is a school library, and not a public library, not an academic library either. Then, you need to figure out what a school library needs to do; and what you need to do as a school librarian.

Then you have to break down the school missions into different layers, for example, what does the headmaster expect; then the administration, the management style, the curricula, and finally the parents.

For example, year 1 teachers are planning to ask the students to develop topics for their school projects by the end of the semester. In order to support their teaching and learning needs, we (librarians) need to order books or other resources beforehand, so that by the time the students are asked to work on their assignments, we will have the materials readily available in the library; and they could just come and check out the materials and bring home to work on their projects.

I know you mentioned about the printed versus online, and I understand that printed books can provide a sense of security, and sometimes the amount of information available online can get overwhelming. And often times, children do not know what to choose. So when these young year 1 and year 2 students come to the library, I could just tell them this is a new (printed) book on dolphins and they are already happy with my recommendations.

What I am trying to say is that you need to find your way to work with the senior management; and figure out the limitations of the school that you are working for because every institution has its limitations which are resources-, staffing-, policy-related, and so forth. For example, because of my previous background in public librarianship, I really wanted to do book fair for this school. I tried proposing this book fair idea to the principal several times; unfortunately, it never got approved because this school is operating under the umbrella of the Po Leung Kuk Charity Organization[4]—hence, there is not much we could do about it, because doing a bookfair would mean inviting the local bookstores to come to our school, and sell books to the whole school community at a discount (for example, 20% discount for teachers, and 10% discount for students, and so forth.) It would no doubt be a mutually beneficial event. Unfortunately, we simply cannot do it at this school because book fairs would involve profitmaking. Eventually, we divided to do the *Book Swap* instead, which involved no money after all; and the principal, the parents, and the students—everyone was very happy with the end results. As it turned out, the students also do not mind reading second-hand books. And we also talked with the students beforehand that those books meant for *Book Swap* should all be in good conditions.

---

[4] Po Leung Kuk—Homepage. Available at: http://www.poleungkuk.org.hk.

With reference to the curriculum—the school that I previously worked for, they followed the General Certificate of Secondary Education (GCSE) curriculum. And library information literacy skills instruction was not part of the GCSE curriculum. The whole idea of information literacy (IL) is to teach students how to search for information; select, evaluate; synchronize information, and most importantly how do you apply it, and so forth. But this set of IL skills is not measured under the GCSE. But of course, they are experiencing the need to introduce IL to GCSE right now. But when I first arrived at that school, they did not believe in any of this IL instruction. So, I just kept pushing IL to the school, but then I felt the resistance amongst the other teachers was very strong. But for this school (CKY), it is relatively easy for them to accept new changes as long as they see my proposals are appropriate and feasible.

To NJ—before coming to Hong Kong, you were working as a public librarian in the United States. With reference to your background and experiences in public librarianship, how does it in anyway contribute to your current work as a school librarian?

NJ: Coming from public librarianship, the training and concepts of servicing the general public are found to be most useful and relevant to my current job as a school librarian. After I started working here (CKY), I have made a few changes in terms of the library regulations/operations. For example, originally the magazines could not be loaned out to students and I immediately asked, *"Why not?!"* In addition, I did not understand why the students were not even allowed to check out the back issues either, then I was told that because the back issues were not cataloged, as they had no barcodes attached.

Then I discussed with my team and we concluded that we could catalog them—just not do full cataloging; rather key in the magazines' main titles and the range of issues that are available. I simply did not see the need to perform full cataloging for these leisure magazines for teenagers, especially when we are not an academic library. All we wanted was to make these magazines conveniently available for the students to borrow from the library at any time.

So we decided to change all that; and two months later, we completed cataloging the entire magazine collection. And now, not only you could search these magazine titles via online public access catalog (OPAC), but you could also check them out of the library. I am sure there are some damages and losses in the process of circulation, but it is still worth doing for the students' benefits in the long run.

Another thing I changed was adding the "Book Reservation Function" to expand the scope of our services—this service would save the students the effort and time from looking for their desired books from the book-shelves. So, I would say most student–friendly library policies are espe-cially the flexible ones that make the library materials more accessible to them and their library experiences more enjoyable. For example, in the past, when the students had unpaid overdue fines, they would not be able borrow books. But right now, we started using this new system—via which, students are able to remotely log into their library accounts and renew the books by themselves, even when the borrowed items are overdue. And even their library accounts show that they have a small overdue fine, they can still continue to borrow books. In short, these are the little things that I learned from my previous job as a public library manager.

The major difference between public and school librarianship is that as a public librarian, a large part of my job was to develop strategies, to attract the general public to come to use the library. Whereas for my current job as a school librarian, much of my time is spent on developing collaborative learning projects with other local schools.

When you work in a school library, you will have a much more consis-tent user group. Of course, individual students would have different ques-tions, interests, preferences, and needs. For example, we have a male student here named Max—he is always the first user to come to the library and says, *"Hey! Ms. Niu...."* After saying "Hi," he would just disap-pear behind the bookshelves. There is another student named Caden—he always comes to the library and looks at Google Earth. His dream career is to fly off somewhere in a jet aircraft. The ultimate advantage of having a group of steady users is that you could spend time to get to know them, and thereby doing extra little things to cater for their individual reading interests and learning needs.

So would you say the major advantages of working with a consistent user group is that you would be able to spend more time to try out different things and ideas to maintain their interests; as well as to observe indi-vidual students' learning progress under an ongoing basis?

NJ: Right, it allows me to monitor the progress that individual students are making. Well, for example, a few days ago, I said to one of the students, *"Oh! You do not know how to renew your books? Come over! Let me show you how to do it."* Then a few days later, he came into the library again, and successfully reserved and renewed the books all by himself. And again, a few days later, he said to me, *"Wow! I found this book, and it is really interesting. But do you know how to find this book and that book in the library as well?"* As you can see, through these little interactions with individual students, I can see they are gradually becoming more and more interested in our library, as we are becoming increasingly "self-dependent."

As a school librarian, what parts of your job you find most satisfying?

TN: I would say learning from other classroom teachers that I have fulfilled my role and I am doing a good job as a school librarian—that makes me really happy. I also think I have maintained a good relationship with the parents, because I am currently overseeing a large team of library volunteers and they are all parents. This is an experience that I never got from my previous workplace. When I first started this parent volunteer program, I only had 40 parent volunteers; but now we have more than 120. The sheer increase in number of parents participating already speaks for itself.

JM: I love connecting with other people through books. I enjoy ordering books that later become popular amongst student groups and books, which support the curriculum. I also love dealing with student librarians. Additionally, I like to involve student librarians in doing something other than just shelving books. For example, when Christmas was approaching, I asked them to create a Christmas book display and bulletin board. Finally, they got to look at the end results, and were all very happy and proud of their involvement/achievement. It is always nice to help them realize that the library can also be a fun place—and not just a place for shelving books.

NJ: I also feel the same as Tammy and Joli. For myself, I feel more like a "bridge"—I am linking together the people, students, parents, visitors, and library student interns—and connect to the right place to find the right resources. There are so many resources out there, regardless of being printed or available online, and I am happiest whenever I could connect people to whatever materials they need or want.

Many school librarians in Hong Kong are reluctant to try out new services and explore new ideas, for the reason of being afraid to make mistakes. Or worry that their new proposals or ideas would not get accepted by the senior management or by the other classroom teachers. But at the same time, they complain that recognitions, supports, and opportunities have not been given to them. Do you understand the reasons behind this difficult situation?

NJ: Yeah! It could be very difficult. Often at times, you only have one or two librarians working at a school; and you could easily feel lonely or even isolated. If you do not have the necessary training or other senior teachers who could guide you—then you are stuck. For example, nowadays a lot of people can drive a car; but at the same time, there are others who still cannot drive. You simply cannot give a car to someone without having taken any driving lessons, and then expect them to know how to steer the wheel in a few weeks. And why and how we are driving is more important.

TN: And I would say this really depends on your experience. I think school librarians have to understand different schools have different cultures, and also different management structures and styles. I cannot speak for other school librarians, but I would say it could be really difficult because if you are spending money for the school—since you are dealing with money, the senior management would have an obligation to check on you more often—ensuring what you are doing for the school library is actually making sense; as well as beneficial for the school community as a whole. It is understandable because managing a budget is a very sensitive matter, especially for most education institutions.

Since there is not a checklist like region-wide or nationwide syllabus to follow, do you think this makes the job of a school librarian much more difficult in comparison to the work of other Mathematics, History, or English language teachers?

NJ: Absolutely. (TN: Yeah! Absolutely.) But I think these all depend on how much you yourself want to do as a school librarian. For example, just like this afternoon we are giving this interview with you. I do not think the principal will come and blame us for not giving any library lessons. At the end, it all comes down to our own professional judgments, our experiences, whether we see the need to give library lessons at this particular time, and how frequently these lessons should be given. In fact, we asked for this extra work—meaning that in the beginning, we had to tell the senior management that library lessons are important for the students, and students should be given these library lessons on a regular basis. Since we asked for this, we have to be prepared to handle the extra workload, as well as the additional responsibilities that come along with the job. Just like what I said earlier, make yourself worthwhile, make yourself important, make yourself visible and gradually you can see great and positive impacts on the students, as well as support from the other teachers and senior management.

NJ: Yes, I totally agree with JM. I think it is sad to see any schools without a library or having a fully trained and qualified school librarian to manage the school library full-time. I should also point out that in addition to having the appropriate qualifications; it still takes many years of training and hands-on experience to become a good school librarian. Since we all have different professional trainings, education, and cultural backgrounds, I am sure everyone could bring something unique and meaningful to their professional practices as school librarians.

There is no doubt that most schools would have much lower number of senior teachers. Maybe only ten out of 200 teachers could eventually get promoted and become senior teachers or subject leaders. Hence, the opportunities for promotion are indeed very limited as well as competitive; and will only be awarded to the individuals who could bring the most to the whole school community. For this reason, many school librarians in Hong Kong are frustrated about their narrow and unpromising career paths—causing them to feel they are not important within the whole school. Although, they have all the required qualifications (teacher diploma, school librarian certificates, etc.)—they are not being recognized. For this reason, some of these school librarians are the burnout teachers—they do not enjoy teaching anymore; they cannot or do not want to face the students anymore; and they do not want to deal with the parents anymore—a combination of all these different factors that lead them to take up the "easy" job as a school librarian. But the job of a school librarian is definitely not easy if you take it seriously—it is not easy at all!

So, are you saying it is also up to the individuals to take the school library job seriously or not?

NJ: Exactly, there are also many satisfied principals, who have retired, but still want to open another school. At the same time, there are other principals who are totally burnt out and frustrated with their own jobs—and do not want to deal with the parents or deal with any management issues.

Yes, this is also true for the school librarian profession. So if you are young, you should be passionate and creative about your work, or at least try to learn to find satisfaction from your work.

How do you develop such constant passion, enthusiasm, as well as satisfaction for your work as a school librarian?

NJ: I think it depends largely on how, where, and when you first entered into librarianship profession, for example, who your coworkers are; who your mentors are; what is your own personality like; whether you are intellectually curious—meaning if you are a good learner and a good listener—it is usually a combination of all these different factors. What I am trying to say is that you can only, *"Lead the horse to the river, but you cannot force the horse to drink from it."*

JM: The nicest thing about being a teacher-librarian is that there is a great deal of openness that comes with the job. While other classroom teachers have to strictly follow a set curriculum, we teacher-librarians get to write our own curriculum and adjust it when necessary. Since we do not have a strict checklist—like syllabus to follow, we teacher-librarians have a lot of freedom to exercise our professional knowledge and skills for the overall welfare of our students.

TN: I think you could become passionate by sharing ideas with other people who are equally passionate about their work as school librarians. I think sharing experiences and ideas is really important, because you could definitely learn successful experiences from other people, as well as not repeating their mistakes. I think I am really lucky because at my previous workplace, I had a very good supervisor who taught me a great deal of skills, for example, collection development. In Hong Kong, it is good that we have a professional association called Association of Librarians in

English Speaking Schools in Hong Kong[5](ALESS)—and many librarians from the local international schools and English as medium instruction (EMI) schools in Hong Kong are active members of ALESS. I encourage school librarians to share their experience and ideas with each other, while sharing you are actually the one who learns a lot.

---

[5] ALESS (Association of Librarians in English Speaking Schools in Hong Kong)—Homepage. Available at: http://aless.wikispaces.com.

Tammy Ng, Teacher Librarian

Joli Moore, Teacher Librarian

Jun Niu, Teacher Librarian

School Library

MODELLING YOUR SCHOOL LIBRARY AFTER STARBUCKS? SUCCESSFUL SCHOOL LIBRARIAN STORIES FROM HONG KONG

GLORIA CHAN

PAOC Ka Chi Secondary School[1], Hong Kong (SAR), China

May I ask what did you major in when you were in college? Have you taught any other subjects prior to working as a teacher-librarian?

At university, I majored in Chinese Language and Literature. My qualification also includes a Diploma in Education. Prior to working in this school, I was teaching in a local primary school in Hong Kong. About ten years ago, when I first started working in this school, the principal (headmaster) encouraged me to try managing the school library. That was around the time when the Hong Kong Education Bureau (EMB)[2] implemented a policy requiring a permanent teacher-librarian to station at all local primary schools. And I simply said, "Yes" to my principal's request. Over time, I gradually developed an interest in school library work. I, therefore, decided to study for the teacher-librarian diploma, while I was working as a schoolteacher at another primary school. In other words, I had practical experience in working as a teacher-librarian prior to coming to the current school.

Based on my previous conversations with other school librarians practicing in Hong Kong, many of them complain that the career prospects of teacher-librarian are rather grim and unpromising. For example,

---

[1] PAOC Ka Chi Secondary School—Homepage. Available at: http://www.kachi.edu.hk/en.

[2] Education Bureau of the Hong Kong SAR Government—Homepage. Available at: http://www.edb.gov.hk/en/.

opportunities for a school librarian to be promoted to the rank of senior teacher or headmaster are in fact very rare—do you agree?

I believe that "promotion" would really depend on a large number of different internal, as well as external factors—that is being at the right place, and at the right time. Whether or not the principal values "voluntary reading" or "reading for pleasure" amongst the students is also an important factor in determining the success of a school librarian's career. Overall though, it also depends on how the school librarian views his/her own job; and whether he/she is hard-working, as well as being able to deliver what is expected out of him/her. When I first began my teaching career many years ago, it did not take me too long to decide to take on the responsibilities of managing the school library because I had developed an unexplainable interest in school library work. The principal said that my background in Chinese language and literature was a great "fit" for the (school librarian) position, so I took up the school librarian position with great enthusiasm, and with almost no hesitation.

Based on our conversation, I can feel that you are really passionate, and are able to find a lot of fulfillments in your work as a school librarian—is that true?

Yes, yes!!! It is because everyone from our school, from senior management to junior teachers, are all great supporters of voluntary reading or reading for pleasure. We have also expanded the physical size of the school library by connecting the original (school) library with the adjacent geography classroom—a clear sign that our school truly celebrates the idea of reading for pleasure. And we were lucky to have received a large amount of financial supports from both inside and outside of our school for making our library expansion project successful.

Even when the whole school truly supports reading, many teachers would say, in reality, the status and roles of a teacher-librarian are sometimes considered inferior to that of a physical education (PE), music, or home economics teacher. The reason is that there are many health and safety regulations concerning the students that these PE and home economics teachers must observe carefully. In the case of teaching music, one would at least need to be able to play the piano, and teach basic music theory.

On the contrary, many principals do not always think that it is necessary to hire a qualified professional to manage the school library full-time—for the reason that the library circulation operations and book displays could be easily managed by someone with minimal training or job-related qualifications. In their opinions, no special training and skills are involved. Even when you have someone who is not so skilled in cataloging (e.g., assigning wrong classification numbers or subject headings to the book items), or not so active in reading promotion, it would only mean that the students might take longer time to retrieve the book, or more students would go to play sports, instead of using the school library—the library itself could still be fully operational, and yet without causing any disruptions to students' academic learning and safety. In many principals' opinions, the school library is seen as an add-on facility for the extracurricular activities, and almost never contributes directly to the core curriculum.

Especially for physical education teachers, they need to know how to handle student injuries. The physical education teacher and music teachers can lead students to participate in different inter-school competitions, thereby bringing recognitions and prestige to the whole school. For such obvious reasons, school principals would often find a teacher close to retirement age or someone who prefers fewer and lighter responsibilities to maintain the school library's daily operations at a minimal level—would you not agree?

Actually, I, myself, do not see it that way. In my honest opinion, "Learning to Read, Reading to Learn" is one of four key areas of the Hong Kong Education Bureau Curriculum Reform is aiming to develop. And under this reform, all teacher-librarians are expected to play a vital and yet, very influential role. According to the reform policies, teacher-librarians are expected to collaborate closely with other classroom teachers—using the school library as the "center base," to develop a wide range of educational activities—with the aim of guiding their students to develop a set of important skills and values that would equip them with the abilities, motivations, and the right attitudes necessary for them to become lifelong learners of the 21st century.

I have to admit that the nature of work of a school librarian can be quite monotonous at times, and there is a lot of truth in it. When the other subject teachers talk, they are aware of what is happening in the different classrooms, but they might not necessarily know what the library is busy working on. For these reasons, it is easy to feel "left out," or unimportant (almost like a junior) staff member of the school. In the beginning, I was kind of afraid to start a conversation with other teachers, let alone approaching them. This is genuinely how I felt during the first five to six years of my work as a school librarian. It got to a point where I realized that if my work went on like this, it would really become a vicious cycle with no way out, and I would be drained of all my energy. I would eventually lose direction and become unhappy with my work, maybe even to the point of leaving this working environment!

In the earlier years, I did try to organize a reading club for the whole school. And, in the beginning, I often felt like I was the only person fighting the whole battle, because I did all the preparation work by myself. But over the years, I gradually discovered that there were other colleagues within the same school, who were just as keen as I am, in terms of creating a culture of voluntary reading for the whole school. For example, I have a teacher colleague who is very skilled in building models. So I asked him to recommend books on building toy models. In other words, I tried to find other like-minded colleagues, spent time organizing the reading clubs with them, and more importantly read together with the students… and over years, we created different book clubs with themes or focuses surrounding different hobbies shared amongst students and teachers. This is how I started to change things. Maybe I lucked out, but things are so much different now, and it is absolutely wonderful.

In addition to sharing with others, another way to overcome the feeling loneliness is to be proactive—that is, to initiate the first step to collaborate with other subject teachers. For example, earlier this year, I organized a major book-buying trip for the whole school. Under the supervision of the teachers, all students from Form 1 to 5 went on a field trip to the local mega bookstore—to select books to be added to our school library's collection. For this kind of large-scale activity, it did require a large number of teachers from all levels to participate. Prior to the book-buying trip, I also had to prepare a lot of information (e.g., guidelines, hands-on manuals, etc.) for training the other teachers—so that their students would not end up purchasing duplicate titles. I also worked hard to ensure that this activity would be loaded with fun, and enjoyable for every student to take part—thereby leaving them a positive experience that would then last over time.

Furthermore, this year I collaborated with the teachers of the Chinese and English Department—as a team, we designed a "Student Passport."

The Student Passport worked like this: the Chinese and English subject teachers wrote down a specific goal and the teachers would stamp the passport after a student had successfully reached that set goal. Some examples of goals would be the number of designated Chinese-language books read by this student within the set period. This was not just a project meant for supporting the Chinese language syllabus, but something that also aimed at developing students' interest in voluntary reading, or reading for pleasure. Of course, they also had goals that were designed for preparing the students to do well in dictations and examinations. After the students had successfully collected a certain number of stamps, they would then receive book vouchers, which would be presented to them during the student award ceremony—with the hope of encouraging them to continue to read and/ or buy more books. Collaborating with the Chinese and English Department really made my work more enjoyable because it made me feel like a valuable member of the whole school community—participating actively in the overall design of the whole activity and feeling that I could actually make a positive difference in students' learning. Because of all these team projects, I did not feel lonely at all. At the end of the day, it all depends on how an individual teacher-librarian carries out his/her own work, and how much effort and creativity they are willing to invest in for realizing his/her own roles, as well as the full pedagogical and creational potentials of the school library.

Despite your own successes, have you heard other colleagues in the community complaining that their schools and the senior management not supporting what they do as school librarians? For example, some principals would often say to them, "Why should I waste the manpower and resources to hire a qualified school librarian, when the (school library) work could be easily carried out by a clerical staff (with minimal training and qualification)? Or would it not be more cost-saving to have the school library managed by another classroom teacher as an extracurricular activity on a part-time basis?"

You mentioned a very important point: the key is how important the school's senior management values the educational and recreational roles of the school library. At my school, the principal places a lot of emphasis on voluntary reading (reading for pleasure) amongst our students. At the same time, I also take a lot of effort to ensure that he understands that the school library is not merely a warehouse for printed books, but is in fact the heart and soul for learning, reading, and recreation for the whole school. Every time someone (from either inside or outside) visits our school, they are invited to visit our school library first—as it is always featured as the showcase of our entire school. Having a school library, with facilities, services, and professional assets that the whole school could take pride in is certainly important for gaining recognitions for being school librarians. When students are willing to read voluntarily, it would not only enhance their curiosity and thirst for learning, and elevate their academic level, but it will also broaden their knowledge and understanding of the world.

Nowadays, education in Hong Kong, taking the subject of Liberal Studies as an example, requires a lot of reading that is outside of scope of regular textbooks. Without question, the school library plays a very important role in this context. In order to assist students with all of their academic studies, this year, I created a bibliography involving the nine core areas of the local high school curriculum, which is: (1) Chinese language, (2) English language, (3) Liberal Studies, (4) Mathematics, (5) Physical Education, (6) Arts, (7) Humanities, (8) Geography, and (9) Science and Technology. We (teachers and school librarians) then worked together as a team and selected books to be added to the school library collection—these are the books aiming for supporting all nine subjects mentioned above.

The (school-library-related) activities mentioned by you above, were these ideas originated by you or by your other teacher colleagues?

In fact, a lot of the ideas for new projects related to the school library came directly from the vice-principal. She has always been very supportive, and without any hesitation, treats our library-based educational programs as one of the school's core academic projects. I have been working very closely with the same vice principal for many years, and she has been giving me full support and encouragement, as long as the library projects are in line with the school's mission.

Apart from managing the school library, are you required to teach other non-library-related subjects? As you understand, some schools or principals are not as supportive, and tend to think that teacher-librarians should take up other classroom-teaching duties, in addition to managing the school library—with the aim of maximizing the human resources available. What is your opinion in this regard?

Currently, I am also required to perform other non-library-related classroom teaching duties—that is, to teach a class of Form 4 students in Chinese language.

Is your nonschool-library-related teaching workload heavy? Would it sometimes affect your work as a school librarian?

Yes, my heavy classroom teaching duties could sometimes affect or interfere with my library work. In addition to working as a school librarian and teaching other non-library-related duties, I am also serving as the vice-chairperson of the Parent-Teacher Association. Because of this, I also need to coordinate other large-scale events, such as graduation ceremonies, as well as opening and closing ceremonies for the whole school.

Do you think it is an advantage or disadvantage for a school librarian to take up other non-library-related classroom teaching duties?

Based on my experience, I would say not having to take up other classroom teaching duties would be better, because the school librarian can devote all his/her time and energies to managing the school library, as well as implementing other educational activities and programs related to reading. Taking today as an example, originally, I planned to spend more time on doing cataloging, but I ended up having to spend the entire morning on supervising students taking exams. Prior to their exams, I also had to spend time on examination preparations. For the subject of Chinese language, there are five major examinations throughout the whole academic year. As you can see, I have to spend a lot of time on performing tasks and duties that are outside the scope of the school library.

Having to take up other nonlibrary-related classroom teaching duties, do you think it may be useful for the professional development for a school librarian in the long run?

As mentioned earlier, I am teaching Chinese language, which in my opinion, is not that helpful. On the contrary, teaching Liberal Studies and running the school library—is undoubtedly a perfect match. On the basis of these two subjects, Chinese language and Liberal Studies, students' overall performances are evaluated based on their research projects or written essays, and not via traditional examinations. Furthermore, the nature and contents of the Liberal Studies course are more geared toward developing individual students' analytical skills and critical thinking— which are really the basis for inquiry-based learning. In other words, students are required to do a lot of voluntary reading (outside the given textbooks) on their own. In order to do well in Liberal Studies, students must develop their own understanding of the issues being discussed. For all such obvious reasons, being able to express themselves well in their own mother tongue (Chinese) is really the key, especially when it comes to writing essays. Thus, in this context, the Chinese language classes are absolutely vital.

Have you ever thought about giving up being a school librarian, and work as a regular classroom teacher, teaching other traditional academic subjects instead—which in turn would give you a better path or career progression?

This thought indeed crossed my mind several times during first ten years of my career (as a school librarian). I have been in this (school librarian) profession for over 12 years now, and interestingly, I have already given up on the idea completely because the longer I have worked as a school librarian, the more satisfying and rewarding I find my job to be.

Back to the key issue we discussed earlier, why do you think some schools are willing to invest so much manpower and resources onto developing the school library—using the school library as the heart and soul for learning and teaching for the entire school community, whereas other schools would merely use it as a warehouse for books or treating it a as the detention cell house for ill-disciplined students?

This all depends on the views and attitudes of the school principal. As you explained earlier, many principals may not see the direct and positive impacts that the school library could have on students' overall performance and development. Without something tangible, measurable, or quantifiable, the school administration may decide not to invest too many resources into developing the school library—resulting in allocating the school librarian to take up other non-library-related teaching duties instead. Or, in the worst case, assigning a clerical staff (without teaching qualification and/or with minimal training) or an unmotivated, soon-to-retire teacher to oversee the school library full-time—with the purpose of keeping the school library's services and functions at a minimal level. This scenario is very common amongst the school library community in Hong Kong. However, our principal simply holds a different view from this.

How do you think a school librarian can change a principal's view toward the values and pedagogical potentials of the school library and the person who is managing it?

As school librarians, we simply have to be proactive in terms of conveying to the principal the values of the school library—by simply stating that reading is the foundation and springboard on which all academic skills of an individual are built upon.

I can feel you are very happy with your work as a school librarian.

Yes, it is because I have genuine support from the whole school. Some people would say that although the current principal is a strong supporter of the school library, it might not necessarily be the case with the next or future principals. I have also heard of other cases that the principal only wishes to treat the school librarian as an administrative staff—with the idea that more work could be done by the school librarian, since all administrative staff are only entitled to 18 days of annual leaves, without the benefits of enjoying the long summer and winter breaks as the other regular teachers. Luckily, it is not the case for our school!

In addition, I am not a person who is afraid of rejections and making mistakes because being afraid to make mistakes would certainly kill your creativity, and kill your motivation to move forward with your goals.

As you can see, our school library has undergone major physical expansion and renovation recently. Without question, this kind of expansion created immeasurable amount of extra administrative work for me personally. During the preparation stage, I also visited more than ten local secondary schools in the same district—because I wanted to see what their newly modeled libraries looked like and learn from the other's successful experiences.

So during the feasibility study stage, I gathered a lot of information from other school librarians who had been through similar library innovation experiences—to collect successful ideas and facts on how an inviting and successful school library should look like and function. They all warned me that this kind of library renovation project would put a great amount of burden as well as stress on the school librarian in charge. However, after weighing all the pros and cons, I still decided to go ahead with the renovation project, because the former school library was already in very poor physical conditions. More importantly, I really disliked the idea of a school library that is physically not unappealing to the students. In my original idea, a large library could attract and accommodate more students. So, I took the first step and explored with the principal the feasibility of connecting the adjoining classroom to the school library, thus expanding the library's physical space. After considering my proposal for a period of time, he finally approved my library expansion proposal, and thought that it would be a good use of the resources available.

Over the years, as part of my professional development as an educator, I have attended several courses on self-improvement and ways to overcome the fear of failure. Through such courses, I have learned that, "All successful people are not afraid to take on challenges, try out new things, or making mistakes. In fact, one will not succeed without failure at one point. If you are not making mistakes, you are simply not making progress!"

It has always been my hope and dream to create a school library with pleasant and attractive interior surroundings that the students would never want to leave. But, this kind of major renovation simply required a lot of money. In order to raise enough money to proceed with the innovation project, I had to pull together my own fundraising campaign. With full support from the principal, I sent as many as 200 letters to major financial groups, institutions, and real estate companies in Hong Kong—to ask for donations of money. During the initial fundraising stage, I also called up my friends, who were successful owners of different commercial businesses, urging them to come to take a look at our school library expansion project—expressing to them our urgent need to raise funds for the school—and actively and persistently pleading for their participation in a joint effort. Although most organizations did not respond, eventually, some did, and one generous financial conglomerate even donated to us a total amount of HKD $200,000 for our library expansion. Eventually, I successfully raised more than HKD $400,000 for the entire renovation project. Through years of experience as a school librarian, I think the key factors for success are that you first need to have sound, realistic goals; at the same time, unafraid of extra hard work to achieve these goals, and not to be so easily discouraged by mistakes and rejections.

It sounds to me that the school principal has a lot of faith and trust in you, and in return gave you a lot of freedom to exercise your professional judgment in terms of how you want run your school library—is it true?

The principal of course looked over all the letters after they were drafted. I once asked him how he wanted the new school library to look like and his reply was only two words: "Enjoy Reading!" As far as our vision/theme goes, we decided to model the school library after the atmosphere of a coffee shop or a commercial bookstore. In fact, many students commented that our school library looks very much like Starbucks.

Compared with other traditional academic subjects, do you think managing the school library work poses more challenges and difficulties (technically speaking)? It is because other academic subjects would have a clear checklist-like course curriculum or syllabus for the teachers to follow. However, the teacher-librarian's work has to be designed and executed solely by the teacher-librarian, from beginning to end. For a majority of the schools, they only have one single teacher-librarian working for the entire school. For this reason, when teacher-librarians encounter problems in their work, it is very difficult for them to find other immediate colleagues or teacher-librarian within their own schools seek advices or to consult your difficulties with—do you agree?

I totally agree. Initially, I had to get used to the idea of working alone most of the time, and to overcome the feeling of loneliness. In the school library, I have to make "solo" decisions on a lot of things. I also need to compete for library class time with other classroom teachers—to ensure that students would acquire the necessary information literacy, reading strategies, and information searching skills via the library lessons that I am teaching. All the above tasks and duties require my direct involvement, because all these tasks and duties fall under the job descriptions of a teacher-librarian. In order for the reading activities to be fun, interesting, and engaging, it really takes a lot of creativity, time, and joint efforts. I also need to be mindful that my recommendations would not get in the way of their teachers' teaching plans, or messing up their schedules. Otherwise, they would worry about the increased workload, or express reluctance to work with you because of the inconvenience caused by your suggestions for trying out new things.

Do you have any other interesting stories that you would like to share with the readers?

Building rapport, taking part actively in assisting other classroom teachers with their teaching duties, and involving teachers from different subject disciplines to initiate different educational activities for engaging the students with their learning are always a good starting point for establishing long-lasting collaborative relationships with other subject teachers, as well as gaining recognitions for what you do as a school librarian. As school librarians, we do not have to be afraid of being overly proactive in terms of marketing our library services—so that the true recreational and pedagogical potentials of the school library could be fully realized.

Gloria Chan
PAOC Ka Chi Secondary School1,
Hong Kong (SAR), China

Interior of the PAOC Ka Chi Secondary School Library

Interior of the PAOC Ka Chi Secondary School Library

CHAPTER 8

ZARAH GAGATIGA: THE FILIPINO SCHOOL LIBRARIAN IN ACTION

ZARAH GAGATIGA

The Beacon Academy[1], Biñan, Laguna, Philippines

Please provide a brief self-introduction and tell us about your professional and educational backgrounds. Could you tell me what you studied at university? Are you a second-career school librarian—meaning that did you have other careers before becoming a school librarian?

I wanted to major in English in college. I went to a teacher training university in the Philippines. I graduated with a Bachelor in Secondary Education with a major in Library Science from the Philippine Normal University (PNU).[2]

This was in 1994, and the library education program has dramatically changed since my time, more than 20 years. With the Board for Librarians in the Philippines[3] and the passing of the law, professionalizing librarians, changes happened—one is the inclusion of IT courses. So, the course has been changed to Bachelor of Science in Library and Information Science (LIS).

The education courses are no more compulsory, and it is the choice of students to minor in education. Therefore, LIS was put under the College of Science and Tech, not humanities or College of Arts and Letters or College of Education. I think this has an effect on how school librarians get their training at the university level, because, while the new LIS program is infused with IT courses, only one course is on school librarianship. General

---

[1] The Beacon Academy—Homepage. Available at: http://www.beaconacademy.ph.

[2] Philippine Normal University—Homepage. Available at: http://www.pnu.edu.ph.

[3] Board for Librarians (Philippines)—Homepage. Available at: http://www.prc.gov.ph/prb/default. aspx?id=22&content=126.

education courses for LIS students do not include prerequisite courses for those who wish to specialize in school librarianship.

It is only now that I realize LIS students who want to pursue school librarianship must have pathways and/or tracking system consisting of courses to take in college to prepare them as school librarians.

I think I was lucky to learn early on that my course is preparatory to being a school librarian. What happened to my dream of teaching English? Well, I took Reading Education and Literature Cognates, which further infused and enriched my education in becoming a school librarian after I graduate. I was also lucky to be employed in schools that are progressive as far as literacy instruction programs are concerned. So, school librarians and libraries play a role in the delivery of the programs, curricula, and pedagogies, too, in these schools.

This is not to say that it has been easy for me as a school librarian. I have worked in progressive schools, but the stigma and stereotype of school librarians as mechanical and auxiliary staff were challenges. Even now, I work in a school offering the International Baccalaureate (IB) program[4], there are teachers in our school who do not wish to work with me as far as instructional functions are concerned. They perceive my work as warehouse type. I keep everything and provide access, but no active involvement in teaching and learning. This is the challenge, but I like it.

You said other teachers in your school who do not wish to work with you as far as instructional functions are concerned—does this negative attitude usually come from the older generation classroom teachers? What do you have to do in order to convince them to collaborate with you?

Instead of the age gap or generation, it is a cultural factor, I think. People who grew up in homes surrounded with books, conversations, art, and varied learning experiences in their education tend to be more aware of how libraries function, and how it can help in a person's overall development.

So, how is collaboration nurtured? There are many ways to do this and in my experience, I have found some strategies as helpful for my work and for teachers. Here is a list:

---

[4] International Baccalaureate—Homepage. Available at: http://www.ibo.org/about-the-ib/.

1. Inclusion of library and technology training during in-service programs especially for new teachers—yes, we need to teach teachers library use, technology use, and integration;
2. Regular library coffee with teachers—where the library's resources, new ones, old ones but are still useful, ones that are not being used and needs to be decided on keeping or weeding out are displayed, talked about, examined, and reviewed;
3. Librarian joins regular faculty meeting to know the cycle of lessons in units and curriculum changes;
4. Librarian follows up with department coordinators the needed resources for teaching and collaboration;
5. Regular library newsletters focused on teachers' needs—resources, pedagogy, and current trends in teaching and learning;
6. Online presence and social media—library rides with the school's webpage and social media accounts.

I keep a positive attitude towards teachers using the library. For those who frequent the library for our services and programs, I encourage them to share with co-teachers what they get from them. I learned over time how powerful these two things are: listening and word of mouth—tools that can boost customer service.

Is the Beacon Academy which you currently work at an international school?

Yes and no! Yes, because the Beacon Academy is an authorized International Baccalaureate School. No, because the school is not chartered as an international school. By law, we are a Department of Education accredited school as well. So, we are a private school, offering the IB (International Baccalaureate) Programs and Filipino in characteristics and culture.

Since the Beacon Academy is an international school—could you tell us about the school-library situations (in terms of staffing, requirements for professional qualifications for school librarians, amount digital resources, collection size, etc.) for a regular local elementary/secondary school in the Philippines? Are all regular local schools in the Philippines all equipped with a school library—that is managed by a full-time school librarian?

There are two scenarios: public school library and private school library. Private school libraries and librarians are challenged to infuse the curriculum with information media literacy, thereby, assuming a more proactive role in teaching and learning. The basic school library functions are performed and many are successful in this area, but it is the integration of school libraries into the practice of teaching and learning pedagogy that many school librarians struggle with. Public schools (in the Philippines) have no libraries since there are no job items in most schools. This is politics—the creation of job items go through many tables, from the Mayor's Office up to Congress. I get the impression that, for this to happen, the Board for Librarians[5] need to help, or the Association of Philippine School Librarians[6] must campaign for awareness, and to advocate the importance of school libraries to stakeholders. The funny thing is, to augment the scarcity of books and reading materials, the Department of Education[7] instituted a library hub back in 2003. After more than ten years, many of the hubs are not functional.

There is this *Standards for Libraries*[8], issued by the Department of Education (Philippines)—it has everything on staffing requirements, professional development, collection standards, and physical space management. Private schools can follow the basics as the budget is financed by paying parents. The public school libraries (in the Philippines) depend on the national budget allocation, which involves a lot of politics. This situation get worse with the overpopulation of students and lack of teachers and instructional materials. This has been the scenario and situation for decades. It is a time to think of creative ways to solve problems and issues in the Philippine school librarianship.

When you said many of the hubs are not functional—why are these library hubs NOT functional?

---

[5] Board for Librarians (Philippines)—Homepage. Available at: http://www.prc.gov.ph/prb/?id=22&content=125.

[6] The Philippines Association of School Librarians—Homepage. Available at: https://paslinews.wordpress.com.

[7] Department of Education, Republic of the Philippines—Homepage. Available at: http://www.deped.gov.ph.

[8] *Standards for Philippines Libraries* (Department of Education, Republic of the Philippines)—Available at: http://www.deped.gov.ph/sites/default/files/order/2011/DO_s2011_56.pdf.

The hubs are not functional primarily because there are not enough librarians managing them. Why? Too few librarians and no job items for librarians.

When you said public schools (in the Philippines) have no libraries since there are no job items in most schools—what does "No Job Items" mean?

This means a legitimate job position issued by the government. If there is a job item, there is budget for salary and allowance. Job items go through the local government unit as a request, to be approved by congress. It is a long process and it can be frustrating.

Could you tell me the social and economic backgrounds from which your students come? In addition, what kinds of work do a majority of their parents do for living?

Our students are children of the elite and the influential in the Philippines—business people, politicians, landowners, the old rich, and so forth. But, we have a 25% scholarship goal, meaning that we admit students from different backgrounds and socioeconomic classes for diversity, interaction, and richness of learning experiences. So far, we have reached 10% of our scholarship goals. We are looking for more (student) scholars.

Choosing a career in school librarianship, was it an active choice out of personal interest? Or it was by chance and circumstance?

It was my mom's choice for me. But, no regrets when I took the hall down the Library Science Department at PNU. I love being a school librarian. I am able to teach reading skills and literature, promote books, and the intelligent use of resources.

Are you currently working as a solo librarian in the whole school?

I am a school librarian with one staff assisting me. We only have 90 students in our school.

Could you describe your typical day at work as a school librarian?

The first hour is routine: housekeeping, checking on staff's work for the week, what has been accomplished, what is in the e-mail that is work related, mostly attending to teachers' requests and students asking for consultations on research.

I am the research coordinator this year. Then, I do cataloging work and reading promotion. I manage acquisition work, but I do not handle account payments. There is also the information management system to manage with my staff. Once a week, I sit in faculty meetings. Twice a month, I plan with teachers who need information literacy classes. But most of the time, I work with students assisting them in research and homework face-to-face or online.

Do you need to take up any classroom teaching duties, in addition to fulfilling your roles as a school librarian?

I teach information literacy lessons, mostly on accessing resources, locating and searching for information, strategies mainly. This is scheduled. Since all subjects have an IL (information literacy) lessons, I pick grade levels to start with and amplify or strengthen as students move up from grades 9 to 12.

As a school librarian in your region, is there a nationwide or region-wide syllabus or curriculum that you need to follow, in terms of performing your work as a school librarian? If not, do you think it is feasible to implement a region-wide syllabus for school librarians? The absence of such a syllabus—do you think it is an advantage or disadvantage?

There is a standard for school librarians and school libraries. It is accredited by the Department of Education, *School Library Guide Order no. 56 s. 2011.*[9] The standards were made by the Board for Librarians. Overall, there is also the law, *Republic Act 9246*, that professionalizes the practice of the profession. It is through this law that a code of ethics for librarians is spelled out, too. So, all librarians in the Philippines need a professional license to practice the profession. This license is renewable every three years. Renewal would require credits of continuing professional education like Master's, PhD, attendance to conferences, publications, conduct of workshops, and so forth.

---

[9] *DO 56, s. 2011—Standards for Philippine Libraries*—Homepage. Available at: http://www.deped. gov.ph/orders/do-56-s-2011.

The challenge in the Philippines is that, while there is a set of school library guidelines, not all schools in public and private agencies implement it. There are many reasons: scarcity of school librarians—there are many schools with no school library and no school librarian; the intellectual ecosystem to grow and develop school libraries is not strong enough; the population of children in schools versus the priority in budget is set for infrastructure; availability of teaching materials and teachers to teach the children.

To answer the question, yes, I think a syllabus will help but implementation is contextual and cultural.

What are the expectations amongst your students, other classroom teachers, and the senior management in the school library, and in you—in the context of supporting the overall learning and teaching, as well as the development of other recreational activities of the whole school?

In my school, which is an IB World school, I am expected to teach research skills, support teaching and learning, and promote reading in all formats and media. I am also expected to follow a high degree of professionalism. That is why I commit to grow professionally and personally, too.

Please give a list of successful library programs (supporting students' overall learning and teaching of other teaching staff) initiated by you as a school librarian?

1. *Reading Passport*—where students are advised of books and resources to read and use for pleasure and light reading;
2. *Author Visits*—where I invite authors to speak about their craft and book technology; why writing can be a meaningful career;
3. *Book Spine Poetry*—where book spines are used for poetry writing;
4. *IL (Information Literacy) Wikispace*—where information literacy skills worksheets, activities and resources are set up and used by students in grades 9 to 12;
5. *Freebie Friday*—a monthly info service to teachers for free apps and e-books;
6. *Tech Tuesday*—a monthly info service and library hacks for teachers to help them facilitate research.

What are the major challenges and difficulties faced by you as a school librarian?

I still feel the stigma that I am perceived as a clerk only.

Which parts of your job as a school librarian did you find most rewarding?

It is rewarding to work with students and teachers when they do research, and we both discover new things and insights. I also enjoy working with them and developing a learning community in the process.

When you said it is rewarding to work with students and teachers as they do research, and you both discover new things and insights—please give detailed examples on new things and insights you discovered via this?

Well, just this morning, I learned about the Sino-Japanese War—that there were two uprisings. Helping a student search for information online, I was able to show her techniques for using keywords and advanced searches in Google and in our database. I often find out that students expect to get immediate information. They see the Internet as a magic lamp. You rub it, and a genie comes out to give you what you wish for. But, it is not like that in academic writing and scholarly work. So I tell them, you need to use your keywords wisely, and read documents of any format in a careful manner. Then, I share with teachers this experience. They get to know attitudes and behavior of their students when searching for information. This informs them what to tweak when they give assignments, especially in research. I have been giving teachers feedback since the start of my work here in the Academy. I see changes in the way research is being taught in the classroom because it spills over in actual practice. When we read the works of our students, we know who researched well and who cut corners. Teachers do their job. I do mine by providing a stronger intellectual structure for research through library services and programs. There are also a few teachers who want to team-teach with me when they cover research skills in their units. I think it needs to be formalized....

The professional knowledge, skills, roles, and other job-related competencies for a school librarian—have they undergone major changes in your region in the last five to ten years? In your opinion, what is the future for school librarians in your region?

The changes are major, like the law, but it needs a healthy structure and system made by people, to create an impact to learners.

The future looks bright, of course, especially that we have shifted to the K-12 program and we are stepping up to face ASEAN (Association of Southeast Asian Nations) integration challenges. But it is a slow and deliberate process of change and making an impact. My concern is that proactive school librarians in the field are very few and the research on school librarianship is limited.

Having a passion for school library work, do you think it is something that is inborn (some people would say it a calling) or it is something that could be developed over experience and exposure?

I believe that there are librarians who stumbled into school librarianship by chance and, with the proper mentoring and meaningful work experiences on the job, that can be pushed on to be successful and happy school librarians. I have met a few librarians whose career choice did not include a career in school librarianship, but have found in it satisfaction. So, they stayed on. As for my own experience, my passion is reading and literature. I suppose these two areas made me stay in school librarianship. I love to read and this love of reading is something I have always wanted to share with children. My love for books led me to push myself to create stories, so school librarianship offered me that space and opportunity to tell stories and to write them. As a school librarian, I was able to understand the book industry and the local and international publishing business. I did not know these things were possible during my college days. I was only sure of two things at the time: I love books and libraries, because it offered me a time and space to read so, I choose to work in the school library.

What kind of attributes does a motivated and successful school librarian always possess?

School librarians need to show a sincere interest to children and young adults. They need to model lifelong learning, curiosity, wonder, imagination, and play. School librarians need to be readers first of all. They need to be service oriented, too.

As a school librarian, do you sometimes feel that you could choose to work very hard or do nothing at all—at the end, you would still get paid the same amount of salary? People are sometimes promoted because of their seniority (only they have been here longer), and not because of how well they do their jobs?

I believe in working hard. In an Asian context, I think, working hard is in our core values. It sometimes defines us—how hard we work. But, over the years, I realize that there is such a thing as working smart—knowing when to say no, managing time and energy better, developing networks and linkages to make work easier and accessible. In my experience, I have been promoted because of the work I do and my integrity. What matters to me is that, at the end of the day, I have helped a person, a student, or teacher love a book, find an answer to a question, lent an ear to listen to his or her concerns, contributed to the creation of smooth systems of thinking and learning.

Throughout your career as a school librarian, did you ever have any regrets or second thoughts?

None :-)

If they were to lay off the school librarian or to close down the school library completely, what kind of impact do you think it would have on the overall learning and recreational activities of the whole school community?

It would be the end of the world—especially to those having no access to books—the digital divide would be widened because really. We have to look at the book as a technology and libraries are valuable access points for this kind of technology. So, if this happens, the right to education and literacy is denied to children and parents, especially the mothers who are expected to read books and tell stories to their children.

School libraries/school librarians and inquiry-based learning—do you think they go always hand in hand? In the school environment, true inquiry-based learning could not be carried out without a proper school library that is managed by a professionally trained school librarian.

Yes! My little knowledge of inquiry learning would be the facilitation of asking questions and thinking through concepts, events, issues that shape one's identity and have an effect in the bigger world. This interaction of question and answer, thinking and conversations occur in reference services, in the facilitation of reading programs, and in the teaching of information literacy skills. Some may say, but these can be done in the classrooms by teachers. Of course! And such learning opportunities are enriched, repetitively so, when it is done outside the classrooms: in laboratories, learning commons, online in the Internet, and so forth. School libraries are laboratories, learning commons and can provide online interaction, too. School libraries are part of a learning community where there existed already an intellectual structure for inquiry-based learning.

Now, the school librarian is part of this community and ecosystem, leading and learning at the same time as everyone else.

Regular classroom teacher versus and school librarian in your region, which one do you think would have a more optimistic and promising career path and career progression?

In my country, both have a promising career path. But, our promotion of the LIS degree is very low. There is a need for librarians, but very few are choosing this degree in college.

Do you have any other interesting stories regarding your professional life as a school librarian that you wish to share with the readers?

I met my college professor in School Librarianship last week while I was grocery shopping. I had to introduce myself. I know she had a tough time recognizing my name and face, but, it does not matter. She hugged me and I hugged her. I had to tell her how she made an impact on my career. She actually gave me a tough time in college, but with the best of intentions that I learned my LIS basics, and stepped up in the character development area of the course work. She opened up an opportunity for me to explore storytelling to children. Meeting her is an affirmation that I am doing something right with my life.

FURTHER READINGS

Zarah Gagatiga — Filipino Librarian: http://filipinolibrarian.blogspot.jp/2005/04/zarah-gagatiga-filipino-librarian.html.

The Beacon Academy Library—Homepage. http://www.beaconacademy.ph/?s=library.

The Zarah Gagatiga Interview: Part 1: http://www.rocketkapre.com/2010/the-zarah-gagatiga-interview-part-1-librarian/Librarian.

Why Picture Books Are Important by Zarah Gagatiga: http://picturebookmonth.com/2013/11/why-picture-books-are-important-by-zarah-gagatiga/.

Zarah Gagatiga: https://about.me/zarah_gagatiga.

Zarah Gagatiga
The Beacon Academy1, Biñan,
Laguna, Philippines

Note from a school library visitor

Instruction time in the school library

CHAPTER 9

A LARGE SCHOOL LIBRARY IN KANCHANABURI, THAILAND RUN BY A PHD CANDIDATE

APINUN SEESUN

Teacher Librarian, Thamakawittayakom School,[1] Kanchanaburi, Thailand

Please provide a brief self-introduction and tell us about your professional and educational backgrounds. Are you a second-career school librarian?

My name is Apinun Seesun, and I work at the Thamakawittayakom School in Kanchanaburi, Thailand. It is one of the biggest secondary schools in Kanchanaburi with 2500 students. In addition to my responsibilities as a teacher librarian, I also teach the Thai language and oversee independent study projects.

I received a Master of Arts in Library and Information Science (LIS) from Chulalongkorn University.[2] I am currently a PhD candidate at Silapakorn University[3] pursuing a degree in Curriculum and Instruction.

Prior to this position, I was the head librarian at the Thamaka Wittayakom School, which is also located in Kanchanaburi. I have worked as a cook and a gardener in the past, so school librarianship is my third career.

What is the average literacy rate amongst population in Thailand, and in particular, your region of Thailand?

---

[1] Thamakawittayakom School—Homepage. Available at: http://www.tmk.ac.th/.

[2] Chulalongkorn University—Homepage. Available at: http://www.chula.ac.th/en/.

[3] Silapakorn University—Homepage. Available at: http://www.su.ac.th/index.php/en/.

The average literacy rate amongst the population in Thailand is 94.1 % (that is, number four of ASEAN[4]).

Is there a law in Thailand for punishing parents for not sending their children to school?

Yes, there is. But the lack of enforcement seriously affects students' opportunity to attend school.

It the Thai national curriculum very much exam-based or on the contrary inquiry-learning-based? If the current elementary and secondary curricula in Thailand are only mean to increase the literacy rate (and to eliminate illiteracy) amongst the general population—what roles do the school library and school librarian play in this context?

The Thai national curriculum is called "The Basic Education Core Curriculum 2008." It is aimed at enhancing the capacity of all learners, who constitute the major force of the country, so as to attain balanced development in all respects—physical strength, knowledge, morality, and so forth. They will fully realize their commitment and responsibilities as Thai citizens as well as members of the world community. Adhering to a democratic form of government under constitutional monarchy, they will be endowed with basic knowledge, and essential skills, and favorable attitude towards further education, livelihood, and lifelong learning. The learner-centered approach is therefore strongly advocated, based on the conviction that all are capable of learning and self-development to their highest potentiality.

Thamakawittayakom School—the school that you are working for, it is a public of private school?

The school that I am working for is a public school.

Please describe the social backgrounds of your students? What do a majority of their parents do for work?

---

[4] ASEAN—Association of Southeast Asian Nations.

Most students are poor and their parents are split up. Most parents work in factories. A minority are farmers. Many parents divorced and students are living with grandparents.

What are the minimal professional qualifications for working as a school librarian in Thailand?

In Thailand, there are no minimum standards. In medium and small schools, all teachers can serve as school librarians at the headmasters' discretion, but most librarians working for large schools have relevant professional qualifications.

Was choosing a career in school librarianship an active choice out of personal interest or was it was by chance and circumstance?

I chose to become a school librarian because it was my dream job. I have always loved reading and wanted to instill that same love of reading in my students and coworkers. One of the biggest motivations for pursuing librarianship was to realize my dream—that is, to develop a properly functioning library fully stocked with books. When I was a student, there was a noticeable lack of books in my own school library, and it just was not a great place to enjoy reading.

In Thailand, is it mandatory for every single public or private school to be equipped with a school library? In addition, is it mandatory for all school libraries to be managed by a professionally qualified school librarian? Or there are cases that a school library is only managed by a regular class-room teacher who oversees the school library as some kind of extracurricular activity?

In my country, I think it is required for every school to be equipped with a school library, and it should be managed by a professionally qualified school librarian. Having said that, one could often find many school libraries in Thailand that are managed by regular classroom teachers only.

Are you currently working as a solo librarian in the whole school?

In my school, I am the head librarian and lucky to have staff working for me. I have 100 youth librarians from the Youth Librarian Club who come in to help manage the day-to-day operations of the school library. There are also ten teachers (in our school), who are interested in developing the school library and assist me in many ways.

Could you describe your typical day at work as a school librarian?

In the mornings, I open the library at 7:00 a.m., and I check books in and out, answer questions from students, and recommend books. I also perform cataloging duties and save all data in OBEC (Office of Basic Education Commission), the library automation system. I perform three to four hours of teaching a day and usually close the library in the evening around 5:00 or 6:00 p.m.

Do you need to take up any classroom teaching duties in addition to fulfilling your role as a school librarian?

In addition to teaching Thai language classes, I also design other creative library-related learning to promote strong reading habits, and literacy skills that are necessary for turning young students into successful learners in the 21st century.

As a school librarian in your region—is there a nationwide or region-wide syllabus or curriculum that you need to follow, in terms of performing your work as a school librarian? If not, do you think it is feasible to implement a region-wide syllabus for school librarians? The absence of such as syllabus—do you think it is an advantage or disadvantage?

The nationwide syllabus provides the core guidelines of my work and the regional syllabus allows me to create and adopt activities to meet the learning and recreational of my students.

What are the expectations amongst your students, other classroom teachers and the senior management in the school library, and in you—in the context of supporting the overall learning and teaching, as well as the development of other recreational activities of the whole school?

My school expects me to harness new technologies to achieve inquiry-based learning and teach my students to master various tools related to information literacy, and to understand "new media" and their impacts on students' learning. I also want to make students' learning and reading experiences fun and engaging.

Please give a list of successful library programs such as supporting students' overall learning and teaching of other teaching staff initiated by you as a school librarian?

As the school librarian, I try to identify new and engaging ways to cultivate a love of reading amongst my students. One of the "fun" activities designed by me is to encourage my students to post photos of their favorite books on social media, with the aim of promoting and sharing what they have read to their peers.

I also want my students to develop an awareness of the current affairs happening in Thailand, as well as around the world. I usually pick a current topic that my students feel strongly about. I would then ask them to seek out news sources—and using that as frame of reference when writing their own viewpoints or commentaries. The students are then encouraged to post their own commentaries on the school's newsboard, thereby allowing them to share their opinions with their peers, as well as facilitating discussions.

I also like to have fun with my students through playing the Millionaire Game. Via this Game, I ask them a variety of questions with the aim of testing their common knowledge. If they could answer my questions correctly, they would be awarded with a prize.

What are the major challenges and difficulties faced by you as a school librarian?

The lack of budget has been the main difficulty, but I always try to be positive and find ways to work around the budgetary barriers. I also have the assistance from the PTA (Parent Teacher Association) and my library staff. I try to look at these difficulties as simply small hurdles that can be easily overcome with cooperation.

Which parts of your job as a school librarian do you find most rewarding?

The most rewarding aspect of my job is being able to instill a love of reading in my students. My students might not be able to find immediate use for the skills that I am teaching them now. But, these skills would certain provide a strong foundation for them to become independent, life-long learners, particularly if they wish to be successful with their future careers.

Would you say that the professional knowledge, skills, roles, and other job-related competencies for a school librarian have undergone major changes in your region in the last five to ten years? In your opinion, what is the future for school librarians in your region?

In Southeast Asia in the past ten years, school librarians have been actively developing their libraries (both collections and facilities), in order to meet the learning and recreational needs of the students in the 21st century. Some librarians have really made information technology in the school library a priority. Other librarians would like to do the same, but again, lack of funding is always the most critical issue. An unfortunate reality is that polit-ical, economic, and social realities have always prevented the development of school libraries in some regions. Within Southeast Asia, I feel that the outlook is still relatively positive for librarians, particularly in Singapore. However, in my own region (Thailand), I feel less optimistic.

Do you think having a passion for school library work can be considered a calling or it is something that can be developed through experience and exposure?

Experience contributes to strong and professional competence, which are necessary to become a successful school librarian in the 21st century. Having said that, having a genuine passion passion and lifelong love of libraries and library work still remains the key.

What kind of attributes does a motivated and successful school librarian always possess?

A successful librarian always has a mind for self-development and constantly looks for new ways to improve their library services to meet the needs of students and staff. They must have the drive to help students develop an interest in reading and literacy skills. I believe your love for library work should come from your heart and care about the users of your library.

As a school librarian, do you sometimes feel that you could choose to work very hard or do nothing at all because in the end, you are still paid the same amount of salary?

As a school librarian, I think I should always work hard for my students in order to help them reach their goals.

Throughout your career as a school librarian, did you ever have any regrets or second thoughts?

No, I have never regretted my career. No matter what kind of library work I am doing, I always devote myself totally myself to finishing the task in hand, and work hard to produce something positive for my students and for my school library.

If the school was to lay off the school librarian or to close down the school library completely, do you think it would have an impact on the students?

Of course it would have major impacts on my students' learning. Libraries and librarians are so necessary for implementing inquiry-based learning and the development of students' literacy skills. Students use the library to access a wealth of knowledge (both within and beyond the school library), and we school librarians are there to assist them in that process.

Do you think that true inquiry-based learning could not be carried out without a proper school library that is managed by a professionally trained school librarian? What do you think are the benefits of inquiry-based learning for students?

Yes, I think true inquiry-based learning could not be carried out without a proper school library that is managed by a professionally-trained school librarian. The work of a librarian is driven by inquiry, and we need to work hard to support students with their own quest for knowledge.

Teachers give students a task or a research project and the students must pursue their own lines of inquiry by seeking evidence and other information support their ideas and academic endeavors. With the assistance and guidance of the school librarian, students are able to familiarize themselves with the various resources at their disposal, such as e-journals and databases. When this learning activity is carried out as a group, students have a greater opportunity to learn how to cooperate and support each other, when it comes to doing research. Inquiry-based learning provides students with the chance to engage in real academic exchange and share opinions in order to achieve better end results.

In regards to regular classroom teacher versus and school librarian in your region, which do you think has a more optimistic and promising career path and career progression?

I think both classroom teachers and school librarians have promising careers. However, teachers tend to have better opportunities to move up the career ladder. In Thailand, teachers in general tend to earn better salaries.

Do you have other interesting stories from your time as a librarian that you would like to share with the readers?

In Thailand, the government adopted an Education Act in 2002 that allowed school librarians to fill in for teachers whenever there is a shortage of teaching staff. The extra workload (on top of managing the school library) has driven some school librarians to leave their (school librarian) positions, and become full-time regular classroom teachers instead. I think it is a shame. In my case, I also need to perform other non-school-library-related classroom teaching duties, but I do not mind. In fact, this has given me more ways to contribute my skills and knowledge to enhance the learning experiences of my students. I am also an active member of the Thai Library Association, which add to my already very busy schedule. But I manage my time wisely, so that I could perform extra duties to meet the needs of the library and my school.

Mrs. Apinun Seesun
Teacher Librarian, Thamakawittayakom
School, Kanchanaburi, Thailand

Library activity, Acting for fun

Student librarians' librarians' club and its activities : library service

CHAPTER 10

SUCCESSFUL SCHOOL LIBRARIAN STORIES FROM THE LAND OF SMILES

WANPEN ASAKIT

Librarian, Ruamrudee International School,[1] *Bangkok, Thailand*

Please provide a brief self-introduction and tell us about your professional and educational backgrounds. What did you study at university? Are you a second-career school librarian—meaning that did you have other careers before becoming a school librarian?

My name is Wanpen Asakit. I graduated in 1995 with a major in Library Science at the Faculty of Education from Srinakharinwirot University[2], Bangkok. Ever since I graduated from university, I was always a librarian.

I am currently working as an assistant librarian in Elizabeth Library (Elementary School Section) in Ruamrudee International School[3] (Preschool—Grade 5). From 1995 to 1997, I worked as a librarian in Lumnamping College at Tak province (Vocational Education). From 1997 to 2002, I worked a Librarian in Kasetsart University[4] Laboratory School International Program (Grades K-12). From 2002 to 2007, I was a Technical Librarian in the Griffith Library (Middle School/High School Section) in Ruamrudee International School (Grades 6 to 12).

Choosing a career in school librarianship, was it an active choice out of personal interest? Or it was by chance and circumstance?

---

[1] Ruamrudee International School—Homepage. Available at: https://www.rism.ac.th/.

[2] Srinakharinwirot University—Homepage. Available at: http://www.swu.ac.th/en/.

[3] Ruamrudee International School—Homepage. Available at: https://www.rism.ac.th/.

[4] Kasetsart University—Homepage. Available at: http://www.ku.ac.th/web2012/index. php?c=adms&m=mainpage1.

Lumnamping College was established in 1995. I had opportunity to be the first librarian in their College Library. It was very challenging for a newly graduated student such as me to set up everything in the library. I have learned many new things and have enjoyed working since then.

Are you currently working as a solo librarian in the whole school?

No, we work as a team. The Libraries and Resources Department supports school curriculum, learning, and teaching from preschool through Grade 12.

The Libraries and Resources Department has two teams, and each team has a teacher-librarian and staff who have specific skills such as technical librarian, library aide, and teacher assistant to perform and organize the library works, exhibitions, and special events as a good teamwork.

The Elizabeth Library Team is working in the Elementary Library and Elementary Curriculum Resource Centre (ECRC) to support Elementary section. The Griffith Library Team is working in the Griffith Library and Textbook Center to support Middle School and High School sections.

Could you describe your typical day at work as a school librarian?

My responsibilities in the school library and ECRC is to assist the teacher-librarian in arranging and managing the effective operation of the library, perform cataloging/editing of new and existing books, organize circulation of library materials, and assist students and teachers in the use of library materials and services in order to ensure smooth operations of the school library and other duties as assigned.

Do you need to take up any classroom teaching duties, in addition to fulfilling your roles as a school librarian?

No, I assist the teacher-librarian in teaching information literacy skills, book selection during class borrowing time, research collaborating with teachers by pulling library resources, setting up research links on the library's home page, searching for Internet sources for research project, etc.

As a school librarian in your region, is there a nationwide or region-wide syllabus or curriculum that you need to follow, in terms of performing your work as a school librarian? If not, do you think it is feasible to implement a region-wide syllabus for school librarians? The absence of such a syllabus—do you think it is an advantage or disadvantage?

We do not have any syllabus or curriculum for our Library, but the American Association of School Librarians (AASL) standards for the 21st century learner are integrated to classroom curricula. The Elizabeth Library offers a flexible schedule for library classes for preschool to grade 5 students to borrow or do research in library. Brief instruction of information literacy and the use of library materials skills will be covered, and they are connected to classroom curricula. The librarian helps teachers plan for research units, sets up research links, gather materials, teaches students on appropriate book selection skills, and teaches a class on reference skills at teachers' request. The collaboration among librarians, teachers, and curriculum admins enhance educational growth of students although there is no regional syllabus.

What are the expectations among your students, other classroom teachers and the senior management in the school library, and in you—in the context of supporting the overall learning and teaching, as well as the development of other recreational activities of the whole school?

The Elizabeth Library always cooperates with everyone in school. We support the school curriculum and activities to provide good learning and teaching in a lively atmosphere. According to the curriculum, there are more than 23,000 titles and monthly circulation is no fewer than 15,000 copies. I expect to see students love reading and being lifelong learners in the future.

Please give a list of successful library programs (supporting students' overall learning and teaching of other teaching staff) initiated by you as a school librarian?

A successful project in Elementary Section is the DRA (Developmental Reading Assessment) program. In collaboration with library staff, classroom teachers, and parents, this program supports individual students to choose books of appropriate reading level. Classroom teachers will send individual DRA test scores to the library three times a year (beginning, middle, and the end of the year). Then, students will be limited to borrowing by their right level and some for own choice. Parents are encouraged to select appropriate books for their kids. Library staff provides a reading level for every book in library and ECRC. This helps students develop and challenge their learning step-by-step.

What are the major challenges and difficulties faced by you as a school librarian?

There are tons of information and various resources. We need to encourage students to learn how to get the appropriate information from the credible sources.

Which parts of your job as a school librarian did you find most rewarding?

To see students love reading and progress to a higher reading level. I am very happy when students share what they have learned or explore new things from the books they read with glittering eyes, so I can see how happy they are.

The professional knowledge, skills, roles, and other job-related competencies for a school librarian—have they undergone major changes in your region in the last five to ten years? In your opinion, what is the future for school librarians in your region?

Nowadays, electronic media are best developed for various (mobile) devices. Clients want to access reliable information, quickly, and conveniently. Technology skill is important for librarian. School librarians need to know how to retrieve information from any platform. Online database, e-book, and one-stop search services are needed so that clients can access to the library anytime anywhere. The librarian community in the region is helpful too, so we can share knowledge, ideas, and experience among school librarians.

Wanpen Asakit
Librarian, Ruamrudee International School,
Bangkok, Thailand

Elizabeth Library Team

Preschoolers love to read

CHAPTER 11

A SCHOOL LIBRARY IN THE WONDROUS KINGDOM OF THAILAND

WILASINEE THEPWONG

Samakkhiwitthayakhom School, Chiangrai, Thailand

Please provide a brief self-introduction and tell us your professional and educational backgrounds? Could you tell me what you studied at university? Are you a second-career school librarian—meaning that you have other careers before becoming a school librarian?

My name is Wilasinee Thepwong. I am a teacher at Samakkhiwitthayakhom School[1], Chiangrai. After graduating with a Bachelor's degree in Library Science from Chiangmai University. I continued to study for a Master's degree in Information Science at Sukhothai Thammathirat University, which is an Open University in Thailand. The following is a list of my professional achievements:

1. Excellent Teacher who promoted and developed Thai Language Usage at High School Level Year 1999 from Thai Language Institute, Academic Department, Ministry of Education.
2. Outstanding Person in Library and Information Science Year 2013 from Thai Library Association.
3. The prize for a teacher who promoted Reading in the project of 13th Nanmeebook Reading Club, 2014.
4. The students' trainer in the competition of encyclopaedia quizzes for Thai Youth under the King Royal Patronage. We received a winning award from Princess Sirindhorn on 23 January, 2016.

---

[1] Samakkhiwitthayakhom School—Homepage. Available at: http://www.samakkhi.ac.th:81/ULIB/

Information about the school: The Samakkhiwitthayakhom School is a very special large school. There are about 3,474 students with six levels—that is Matthayom 1 to 3 (Grade 7–9) and Matthayom 4 to 6 (Grade 10–12). For Grades 7 to 9, we have 12 classes for each level, so we have 36 classes. For Grades 10 to 12, we have 15 classes for each level, so we have 45 classes. Most of our students live with their parents, who bring them to school by car or motorcycle. If they live far away from the school, they will come to school by bus or van. Furthermore, those who live more than 60 km stay in a dormitory close to the school. The average income per head here is around 250,000 baht (roughly 1USD=31THB) per year. Most parents are government officials or middle classes.

Choosing a career in school librarianship, was it an active choice out of personal interest? Or it was by chance and circumstance?

I am a secondary school librarian, and I took up a career in school librarianship as my personal choice.

In your country, is it mandatory for every single public or private school to be equipped with a school library? In addition, is it mandatory for all school libraries to be managed by a professionally qualified school librarian? Or there are cases that a school library is only managed by a regular classroom teacher, who is overseeing the school library as some kind of extracurricular activity?

Yes, in Thailand each private and government school must have a library for students, teachers, and other members of the school community. Most secondary school libraries are managed by professional librarians, but in small schools, the teachers who teach the Thai language also look after the school library. Most primary school libraries are managed by regular teachers because there are not enough professional school librarians to do this work.

Are you currently working as a solo librarian in the whole school?

No, I am not. There are two librarians in my school, and we also have four clerks who work in our school library.

Could you describe your typical day at work as a school librarian?

As a secondary school librarian, I also have a teaching schedule for six periods a week. I teach "Introduction to Library Use" or Information Literacy. Other work includes cataloging, working on some documents, entering book data, service work for both teachers and students, circulation, and so forth.

As a school librarian in your region, is there a nationwide or region-wide syllabus or curriculum that you need to follow, in terms of performing your work as a school librarian? If not, do you think it is feasible to implement a region-wide syllabus for school librarians? The absence of such a syllabus—do you think it is an advantage or disadvantage?

Yes! There is a standard curriculum set out by the Ministry of Education. These guidelines are reviewed every five years. There are also a set of professional standards from the office of the Basic Education Commission.

The advantage is that the school librarians are able to work on and set their primary missions easily.

The disadvantage is that for school librarians from some schools, or from small-size schools that lack budget, they are not able to meet all the standard requirements.

What are the expectations amongst your students, other classroom teachers and the senior management in the school library, and in you—in the context of supporting the overall learning and teaching, as well as the development of other recreational activities of the whole school?

My expectations for my students are as follows: I would like every student to know how to search for knowledge—that is not only able to conduct searching online but also have the ability to identify which information is valid, relevant, and correct after searching. They also need to analyze and evaluate information found.

My expectations for teachers are as follows: I would like them to prepare their lesson plans by spending their time researching knowledge from books or textbooks in the library before they give suggestions or introduce the sources to students.

Please give a list of successful library programs (supporting students'
overall learning and teaching of other teaching staff) initiated by you as
a school librarian?

This success began from my idea and plan as follows: I let both our teachers
and students give me a list of books, which they wanted to research and
read. Next, I sent the list of chosen books to a bookstore. I asked teachers
and students to go to a bookstore to choose or select books according to
their needs. I contacted the bookstore in order to procure the chosen books,
according to the government instructions. After procurement had taken
place, I cataloged and classified the books and got them ready for library
service. When everything was ready, I made an announcement to the entire
school community, inviting all concerned parties to start using our library.
In this way, teachers and students are encouraged to participate in book
selection and building the collection for the school library. By doing so,
they are all proud and happy.

What are the major challenges and difficulties faced by you as a school
librarian?

Organizing activities to promote voluntary reading for all students
throughout the school is challenging. It is important to raise awareness of
the importance of good reading habits, which are sustainable in the long
run.

Which parts of your job as a school librarian did you find most rewarding?

When I see teachers and students going to the school library
whenever they have free time to borrow books to read at home,
I feel that my work as the school librarian is totally worthwhile.
The school librarian also has additional roles in collaborating with
subject teachers in order to encourage students to learn, and to look
up the information independently.

The professional knowledge, skills, roles, and other job-related competen-
cies for a school librarian—have they undergone major changes in your
region in the last five to ten years? In your opinion, what is the future for
school librarians in your region?

In the future, much of out duties as school librarians has to do with information technologies, particularly in using computers for managing the school library.

Having a passion for school library work—do you think it is something that is inborn (some people would say it is a calling) or it is something that could be developed through experience and exposure?

I have always had this fascination with school library work, and I think my passion toward school librarianship began and gradually developed since I started working as a teacher at Lampang Kallayanee, Lampang Province. At that time, the school library was not well-developed because there was no teacher librarian working there at that time. Where I moved to Samakkhiwitthayakhom, I was supported by the administrative team, where I was assigned to develop the electronic school library. This means that more services and conveniences could be provided to library users. Moreover, users of ULibM can research book information through the Internet and RFID (Radio Frequency Identification System), which is used for facilitating the borrowing and returning of books. With the convenience brought by such technologies, and improvements of services in the school library, users are able to access to read, borrow, and return electronic books through smartphones.

What kind of attributes does a motivated and successful school librarian always possess?

Diligence, tolerance, commitment, and friendliness to other teachers in school.

As a school librarian, do you sometimes feel that you could choose to work very hard or do nothing at all and at the end, you would still get paid the same amount of salary? People are sometimes promoted because of their seniority (only they have been here longer), and not because of how well they do their jobs?

Yes, it seems like that. When I first started my work around ten to fifteen years back, I felt that I was working hard, but was very depressed when I did not get promotion as others did. At that time, because I was disappointed,

I did not want to do much. Finally, I decided that my responsibility should come first, and I stopped thinking about it.

Throughout your career as a school librarian, did you ever have any regrets or second thoughts?

Yes, as I said I felt disappointed when I worked very hard and nobody saw what I had done. I resigned from the Head of School Library but was still working as a librarian, who did some service work in the library and taught some classes. The new Head of School Library did not have much knowledge about technology and neglected it until some of the library manuals, which I had developed for a number of years, were destroyed. Finally, the administrative team reinstated me as the Head of School Library again. I had to develop the library system so that it could be used effectively. Truthfully, time had been wasted. This made me very sad. Now I understand that nothing is more important than my work. Whether or not I receive a promotion is unimportant: developing our work is more important.

If they were to lay off the school librarian or to close down the school library completely, what kind of impact do you think it would have on the overall learning and recreational needs of the whole school community?

I definitely believe that people who are successful in their careers because they have always been good readers and can effectively make use of the knowledge and other information resources found in the library. If the library facilities were not available, it would mean that the teachers would have to be highly knowledgeable people, in order to carry out the same kinds of tasks that were originally done by school librarians. This is impossible, as they need access to accurate and very reliable information resources.

Although my school library provides various electronic materials, not all students can make good use of these services, because there are still disparities and differences in educational opportunities.

School libraries/school librarians and inquiry-based learning—do you think they go always hand in hand? In the school environment, true inquiry-based learning could not be carried out without a proper school library that is managed by a professionally trained school librarian?

Yes, I believe that the school library, the school librarian, and inquiry-based learning have to go together. The teachers play important roles in persuading the students to go to the library and realize the importance of education and self-studying. It is important to have a school library to be managed by a professional school librarian who provides books, textbooks, or media for learning that are relevant to the students' needs and interests. This undoubtedly encourages children to become avid voluntary readers and constantly searching for the right answers—qualities that are required to become autonomous learners in the long run.

Regular classroom teacher versus school librarian in your region, which one do you think would have a more optimistic and promising career path and career progression?

I think classroom teachers could easily make much more progress than the teacher librarians because the teachers have only one duty—teaching. On the other hand, the school librarian has dual responsibilities—that is to teach and also to work in the library. My school library is open for the whole day. If I want to make progress in my career, I have to work longer, but sometimes I am too tired to do so. Another reason teacher-librarians are the minority when compared with subject teachers, so it is rather difficult to make progress in our careers. Nevertheless, some librarians can make progress if the administrative team (the school's senior management) values the work done by the school librarian.

Are there any other interesting stories that you would like to share with the readers?

Yes, in almost 30 years of my working experiences as a school librarian I can conclude that the students who are successful in their lives—many of them were able continue their education at good universities, and almost all of them are good and avid readers or even researchers. If you want to be successful, you first need to be an active learner. In order to be an active learner, you first need to be an avid reader. Eventually, you will find a way to apply whatever knowledge or skills you gained in your daily life from the joy of reading. You can choose what you want and finally you can apply some of that knowledge in your daily life.

Wilasinee Thepwong
Samakkhiwitthayakhom School,
Chiangrai, Thailand

Self-check RFID (Radio Frequency Identification System) circulation system

Winner of The Reading Encyclopedia Competition, Princess of Sirinthon's prize,
23 January 2016

THE HIDDEN TRUTH ABOUT THE UNDYING DETERMINATION OF A SUCCESSFUL SCHOOL LIBRARIAN

SHARADA PANDEY SIWAKOTI

Chairperson, School Library Consultant, Nepalese Association of School Librarians (NASL), Kathmandu, Nepal

Please provide a self-introduction by telling us about your professional and educational backgrounds. What did you study at university? Are you a second-career school librarian—meaning that did you have other careers before becoming a school librarian?

My name is Sharada Pandey Siwakoti. With reference to my qualifications, I have a Bachelor's degree in Economics and Culture, a Bachelor's of education degree (in English, Management, and Administration), and a Master's degree in Library and Information Science (MLIS). My professional affiliations are as follows:

- Founder Principal at Kshitiz Higher Secondary School
- Founder Director at Mitra Memorial High School
- Founder Central Secretary at Private and Boarding School's Organization (PABSON),
- Former executive member at Government Teacher Association
- Member of Himalayan Lions Club and Chairperson of Sankalpa Lioness Club
- Vice-chairman and now advisor at Matri Nepal Orphan and Disable Organization, Advisor at Rural Women Networks Nepal (RUWON) and so forth.

Could you please tell me what is your official job title? Are you a full-time school librarian or teacher-librarian?

I am a school library consultant. I do consultancy work for school libraries. I give free consultancy to the local government schools, as they are not able to pay, and I charge the private schools nominal fees for the consultancy service that I provide. I also work two hours daily for my small primary school library.

I am also a social worker. Since I was 19 years old, I started doing social work by bringing street children to my home and provided them with basic education. Unfortunately, I could not continue their education as they ran away after a few months. I brought them back twice, but I could not stop them from running away. I also provided education to orphaned and underprivileged children in my own school. I also provided literacy programs to underprivileged and illiterate mothers as well.

The school revenue was used to educate children who are in need of help; we provided many scholarships to children who were in need of help. I am proud that my school has accommodated students who could pay the tuition fees and the school's surplus could be used to cover the cost of education of children who are in need.

You said you are a social worker—I am a little confused—are you currently working as a school librarian or social worker?

I work as a school librarian and also as a principal of the school. However, I devote much more of my time to the developments of school libraries in Nepal. My social work involves education and library development. Every month, I go to teach and support school librarians. I am the president of the Nepalese Association of School Librarians (NASL). This is also considered social work. Nepalese Association of School Librarians focuses primarily on the development of school libraries and emphasizes on developing different strategic plans, educational programs, and activities that are devoted to cultivating good reading habits amongst the local children on a national level. I have also been working for women and underprivileged children for my whole life by teaching them basic information literacy skills and how to access information they desperately need for their daily life.

For example: I teach them how to read and write, how to develop good reading habits, health literacy, law literacy related to domestic violence, women's rights, children's rights, and so forth.

Recently, I was nominated to become a member of the National Martyrs Peace Park by The Ministry of Peace and Reconstruction Ministry, a master plan of the local government. In this program, I will be taking a leadership role in establishing information and documentation centers and e-villages inside the park, and I will mobilize some of the friends to get involved in this project for carrying out the centenary movement in the community. I am also providing books to the local community by helping them establish the community library for local people and children. My group also trains people to become library para-professionals.

Your experience in social work, how does it contribute to your current work as a school librarian?

My experience in social work contributes a lot to my current work through my networks to support library. They are as follows:

- Collecting books from different publishers, writers, personal friends, NGOs, INGOs, embassy, and so forth
- Storytelling through writers, talking about books to children from writers.
- Volunteering works (carried out by my volunteers) for training, monitoring, and supervising school library work in different districts.
- My hardware and software engineering friends support the installation of computers and other library software to those schools that are ready to use Information and Communications Technology (ICT) in their school libraries, at low cost and providing assistance in repairing hardware and software.
- My volunteers and friends of NASL are helping us train school librarians in different districts and remote area with nominal costs.
- Providing supports to the local schools, and to establish library training programs for teacher-librarians through my Association.
- All the stakeholders are involved in holding workshops, awareness programs, and seminars related to school libraries, libraries, education, and so forth.
- Getting help to organize school libraries and other logistic supports.
- Organize advocacy programs like bicycle events, rallies, different curricular, and cocurricular activities.

- Providing opportunities to deaf children to study in private school by integrating normal children with deaf children in the same class with the help of sign-language interpreters.

My social work experience has helped me establish library committees in different districts of Nepal with the aim to sustain the development of school library programs in the long run. We are working for a total number of 21 districts in Nepal. We have already supplied books and helped develop school library programs from both urban and rural areas of Nepal and support many children with their education, particularly in the promotion of good reading habits. I like to give you one example of a Japanese man who is actively involved in our program.

The program officer of UNICEF at the time, named Masahiro Mark Ono, heard about my program and helped me expand my program by providing some external funds. He supported children with difficulties, deaf children, street children, etc. Most of these children have now grow up, and were able to find good jobs in Nepal.

At which school are you currently working as a school librarian? Could you please give me the full name of the school that you are working for?

Our school is called Lok Darshan Primary School (Kshitiz School), and it is located in Kathmandu, the capital and largest municipality of Nepal. Nowadays, I am working as a leader to support the development of school libraries, and provide training to teacher-librarians. I have lost my previous big school due to a land case. I am now working for a very small school and would like to expand its size, but it could not be expanded due to the current bleak economic conditions.

About my previous school—I have run that school for 27 years. I lost that one due to Nepal's unhealthy legal system, and the local land mafia. I was deceived in a land acquisition case by a land broker via buying a piece of land for a school building 12 years ago. I filed a case to the court of law, but I suffered a lot because that land belonged in the name of the temple (Hindu God Ram Chandra) so I was not eligible to use the land anymore.

The students at your school, what kind of social and economic backgrounds do they come from? What do a majority of their parents do for work?

My school is always for ordinary people, of different ethnic backgrounds, and some are of low caste, all mixed, because they cannot pay the high tuition fees, cannot send their children to expensive boarding schools, and also cannot afford to go overseas to further their studies.

Is there a law in Nepal for punishing parents who do NOT send their children to school?

There is no law for punishing parents who do not send their children to school.

Could you tell me what is the overall literacy rate of the general population in Nepal?

60.9% (male 75.1% and female 46.71%)

Why is the overall literacy rate in Nepal so low?

Poverty is the main cause behind low literacy rate in Nepal. Industrial outputs are low by modern standards, and there is no modern scientific agricultural system as well. It is gradually improving partly due the recent compulsory schooling program. But there are lots of dropouts because the children have to support their families' livelihoods by doing different jobs, early marriages, low income, and so forth.

Can you tell me what the economy of the whole country of Nepal is based on?

Agriculture is the main sources of Nepal's economy:
    Agriculture 70%
    Service 18%
    Industry 7%
With reference to Nepal's GDP:
    Agriculture 36.8
    Service 48.7
    Industry 14.5
The economy of the entire Nepal is based on agriculture. 70% of the whole population are in agriculture. GDP is built mainly on remittance of foreign workers.

Nepal's principal economy providing livelihood for more than 65.5% of the entire population, and occupied 37% GDP. Next, GDP is dependent on the remittance of foreign workers, which amount as much as 22 to 25%.

Our local industries are mainly involved in the processing of agricultural products including grains, pulses, jute sugarcane, and tobacco. Rice and wheat are the main food crops of Nepal.

Nepal is a landlocked country with 68% of the total land covered by hills. Only about 20% of the total land is suitable for farming, another 33% are forests, and the rest are mostly mountains. Lowland Tarai regions produce agricultural surpluses, but there is food deficiency in the hills.

The roles of libraries in public schools—since the overall literacy rate in Nepal is still comparatively low—what roles do the school librarians and the school libraries play in the education context? Why would the school librarians and the Education Bureau still insist on implementing inquiry-based learning (which would require a lot more manpower, time, and resources), instead of just investing more resources into teaching the children how to read and write?

Although the overall literacy rate in Nepal is relatively low, we have a large number of schools and school-aged children. If we start teaching these young children information literacy skills at an early age, I think we can really help them broaden their knowledge, and hopefully they could become active and independent learners, rather than relying merely on rote-memorization learning. Although the local education system has made is mandatory for students to do project work for most academic subjects, but nothing is really done on an inquiry-based level.

For this reason, NASL is under the process of convincing the local Government to develop school libraries—that is turning them into centers for supporting true inquiry-based learning for the school community as a whole. If we do not start doing something soon, Nepal will be seriously lacking behind in many fields (in education) in the future. The local Education Bureau is currently focusing on cooperating with all the stakeholders—that is to develop a nation-wide policy for the development of school librarianship in Nepal. The Library Coordination Section is actually relying on me to seek out researchers and scholars specializing in the field of school librarianship (from different parts of the world)—to pull together a seminar—with the hope of identify effective strategies and solutions for the future development of school librarianship in our country. The Ministry has also requested to me to identify potential sponsors, in order to make this seminar happen.

Below is some basic statistics about the local education system, and the overall population of Nepal—so you can understand why we are asking the local government to integrate the school library into whole education system—for the purpose of achieving quality education, and for the aim of implementing true inquiry-based learning. There are totally 34,361 schools in Nepal from grades 1 to 12 (for children aged six to eighteen), and 34,622 ECED/PPC (early childhood centers and pre-primary centers). ECED provides education for children (from two to five years old).

Number of schools by eco belt (in units and level):

Mountain—4,176
Hill—17,339
Valley—2,213
Tarai—10,633
Total—34,361

We have 28 million people of different castes and ethnic groups, with 123 different languages spoken. Nepali is the national language. Some English language programs have been introduced to the local public (government) schools as well. There are 1.5 million elderly people in Nepal, which constitute 6.5 % of the total population in the country. 41 % of the total populations are under age 16. 87.4 % of children are admitted to school at primary level. Among all the school-aged children 48 % are girls.

Student enrollments at primary level by Eco belts is as follows:

Mountain—424,973
Hills—1,849,195
Valley—272,975
Tarai—2,235,742
Total—4,782,885

(Out of that: girls—2,411,849; boys—2,371,036; total—4,782,885)

Student enrollments at lower secondary:

Mountain—145,894
Hills—759,938
Valley—149,125
Tarai—1,812,680
Total—1,812,680
Girls—914,909

Boys—897,771
Total—1,812,680
Student enrollments at secondary school:
    Mountain—64,548
    Hill—352,841
    Valley—84,226
    Tarai—346,954
    Total—848,569
    Girls—421,856
    Boys—848,569
    Total—1,270,425
Student enrollments in higher secondary
    Girls—180,977
    Boys—172,361
    Total—353,338

- In Nepal, students go to school for six years on average.
- About 22.0% of the students dropped out of school.

The following are some of the main causes behind young people dropping out of school in Nepal:

- To support families by performing household chores, (6.5%) because of various financial reasons.
- Early marriage (17.2%)
- Inadequate supports from their own families (7.4%)
- Teacher-student ratio is very high, for example, 36% for primary and 35% for local secondary schools.

In Nepal, around 30–35 % of schools have libraries, and some are well-equipped, but some school libraries only have basic facilities, and many school libraries have nothing more than just books. Many (60%) school libraries have only small book corners. The amount of funding provided for developing the school libraries is about 50,000 Rs. (Nepalese rupees—that is, around US$500). This amount is very small, and also, the book corners are not in proper use due to a lack of school librarians on staff at individual schools.

Since there are no formally trained school librarians, it is very difficult to implement any effective literacy programs for developing children's reading and writing skills. In other words, all school libraries need to be run by properly trained school librarians. In order to do a good job, all school librarians needs training in school librarianship.

Actually, the local government has been advocating this "One School, One Library" policy in 2013, that is based on the advice given by NASL since 1997.

The role of library in the educational context in Nepal can be summarized as follows:

- The school library helps to impart knowledge to the students, teachers, and administrators. In our country, the low literacy rate and the poor quality of the local education go hand in hand. Hence, there is a real need for school libraries and school librarians.
- As students advance through different grade levels, their styles, modes and needs of learning also change. Effective learning could only be achieved if students themselves become active and effective users of the school library. However, the end results (learning outcomes) depend very much on the individual libraries' facilities and human resources (school librarians) available. Unquestionably, school libraries and school librarians both play very important roles in helping students develop lifelong reading habits.
- In short, the school library is not just an information center or a warehouse of books, but also a center for active/independent learning. It is also where students develop the appropriate cognitive skills for making sound (ethical) judgments.

Since the overall literacy rate is still very low—for children who do not get to go to school or have to leave school early—what kind of work would a majority of them end up?

Most of them (both men and women) leave Nepal, and travel to different foreign countries along the Persian Gulf to take up employment, mostly heavy labor work. Whereas for women and girls, a majority of them serve as domestic helpers. Some girls would get married early and became housewives at a young age. Some men do heavy labor work (agriculture work) in Nepal. Some married women also leave Nepal and went to other countries along the Persian Gulf to seek employment (namely: Oman,

Kuwait, Iran, Bahrain, Qatar, Saudi Arabia, United Arab Emirates, and Yemen), Malaysia, and India.

Choosing a career in school librarianship, was it an active choice out of personal interest? Or it was by chance and circumstance?

It was my active choice, based on personal interest. In 1990, I met this Danish Lady, named Ms. Hanne, who worked as an education advisor in Nepal. When we met, we had a series of meaningful discussions about quality education and importance of libraries in Nepal. In 1995, I was invited to take part in a six-months-long educational tour by the Danish Association of School Librarians. Having worked with the team and seen the Danish education system, I was very impressed. Since then, I have continued to advocate for quality education and the importance of libraries in my country. Since then and for many decades, I have dedicated my service to school libraries and school librarians.

Library Science really captured my interest, so I decided to go back to university for my MLIS degree. Earlier, I used to give teacher training to kindergarten and primary school teachers. Later, I started training high school teachers and help them develop their school libraries by making use of waste and second-hand materials (e.g., picture books with simple texts, scrap books, wooden materials, plywood pieces from the furniture factories, coffee beans, seeds from fruits, cereals, paper plates, old books, magazines, etc.) and developed libraries for different preschools. I have also developed a model library for children (from age three to six) by collecting books from different publishers—to establish a small library just for developing children's good reading habits, and also organized different library-related educational events. I continued to share the 'know-how,' by giving various kinds of training to practicing school librarians, and telling different people about the importance of school libraries.

In your country, is it mandatory for every single public or private school to be equipped with a school library? In addition, is it mandatory for all school libraries to be managed by a professionally qualified school librarian? Or there are cases that a school library is only managed by a regular classroom teacher, who is overseeing the school library as some kind of extracurricular activity?

It is not mandatory for the local schools in Nepal to be equipped with a library. Having said that, since 2014, the local government has been advocating this new policy or 'slogan' entitled, *One School, One Library*—a recommendation that has been put forward by NASL[1] since 1997.

Recently, the local government has introduced this policy, that is, all upper secondary schools (grades 11 or 12) should be equipped with a school library. However, this policy has not yet been fully implemented on a nationwide level. Standard requirements and professional skills required for the school librarians to manage this 'mandatory school library' have not yet been fully developed.

In other words, only very few large private schools could afford to have 'professional' school librarians on staff, while many of them have not undertaken any formal training in school librarianship. In short, a majority of the local schools would assign regular classroom teachers to oversee their school libraries as some kind of extra-curricular activity.

Could you describe your typical day at work?

I am not a full-time school librarian. However, I have devoted my professional life to the development of school librarianship for Nepal at a nationwide level, by organizing various trainings, supporting books, under an ongoing basis.

There has never been a strong reading culture and tradition amongst children and teachers in Nepal. Furthermore, the local standardized exam-based education system has also resulted in an unhealthy emphasis on rote-memorization-style learning. We have been working hard to change this situation.

As a school librarian in your region, is there a nationwide or region-wide syllabus or curriculum that you need to follow, in terms of performing your work as a school librarian? If not, do you think it is feasible to implement a region-wide syllabus for school librarians? The absence of such a syllabus—do you think it is an advantage or disadvantage?

There is no government-prescribed syllabus or curriculum either nationwide or region-wide to follow for school librarians.

---

[1] Nepalese Association of School Librarians (NASL)—Homepage. Available at: http://nasl.org.np/.

Yes, we can use the syllabus or curriculum as a set of general guidelines for the school librarians to follow. The NASL has prepared its own syllabus and curriculum for school library trainings ranging from two days up to 35 days long, which incorporates different topics, with formats ranging from seminars, workshops, interaction programs, and so forth. The absence of the syllabus is certainly a disadvantage, but we are coordinating with the local government for developing such a school library curriculum.

Please give a list of successful library programs (supporting students' overall learning and teaching of other teaching staff) initiated by you as a school librarian?

The list of successful library programs is as below:

- Organize library class clubs—establish many reading programs and other recreational activities such as debate competitions, public speaking, book sharing, book reviews, storytelling, drawings, arts and crafts, drama, music, and so forth.
- Organize outdoor picnics—to raise funds and establish new libraries for the local community schools.
- Trainings and seminars—conduct school assembles through the student library clubs on a weekly basis.
- Organize the "Read to Principal" and "Read to Chairman" programs.
- Library visitation programs.
- Celebrate Library Day and School Library Day.
- Develop the experience-sharing programs by inviting renowned scholars and children's book writers to serve as guest speakers.
- Develop "Creative Corners" for children for fostering creative learning activities amongst children, with the ultimate goal of cultivating good reading habits.
- Increase the amount of book exchange and sharing culture.

Which parts of your job as a school librarian did you find most rewarding?

Satisfaction from children—fostering their creativity, and developing their full potentials, by turning them into independent and active learners and keen readers. I am saying this because on one occasion, there was this group of children who presented me with this lovely thoughtful gift—a handmade bookmark, with interesting and meaningful quotations based on school libraries. It was that special moment that I knew I have done the right thing.

The professional knowledge, skills, roles, and other job-related competencies for a school librarian—have they undergone major changes in your region in the last five to ten years? In your opinion, what is the future for school librarians in your region?

Yes, they have undergone major changes. The future of the school librarianship and school librarians in Nepal will be bright because all the stakeholders are beginning to realize the importance of school libraries in the educational context as a whole. And voluntary reading is an indispensable part of any child's learning. The government has also started working on improving the current situation, and has become more willing in listening to us.

Having a passion for school library work—do you think it is something that is inborn (some people would say it a calling) or it is something that could be developed over experience and exposure?

In my opinion, I think it is a combination of both—some qualities are inborn, while other qualities could be gradually developed through time, training, experiences, and exposure, etc.

What kind of attributes does a motivated and successful school librarian always possess?

A successful and motivated school librarian is usually very competent technically; should always be enthusiastic about his/her work, sensitive, scholarly and open-minded, helpful, service-oriented, having a strong sense of responsibility, a cheerful personality, and a good planner...

Throughout your career, did you ever have any regrets or second thoughts?

No, I do not have any regrets. I am happy doing what I do. I have been neglected many times by many stakeholders in the past. I neither quit my work nor my passion toward school librarianship has diminished before of it. Even when my colleagues were expressing high-level of frustrations because of certain situations, I have only become even more determined. I was not able to acquire any funding via the Nepalese Association of School Librarians, except for the Danida fund. The lack of financial supports is the biggest challenge that I am currently facing. Because of that, I am unable to hire a full-time staff member—that is preventing me from continuing my work in advocacy and organizing programs for training more school librarians.

If they were to lay off the school librarian or close down the school library completely, what kind of impact do you think it would have on the overall learning and recreational needs of the whole school community?

It could have major and negative impacts on students' learning and other recreational activities as a whole, because the school library is regarded as the center where all-around development would take place.

Regular classroom teacher versus and school librarian in your region, which one do you think would have a more optimistic and promising career path and career progression?

A teacher in Nepal would have a more optimistic and promising career because it is not mandatory for the local schools to have full-time librarians onstaff.

Are there any other interesting stories that you would like to share with the readers?

There are many, but the most interesting story is the library work carried out by us after earthquake, 25th April 2015.

After this major earthquake, there was a big problem. So we developed a temporary library in an earthquake rehabilitation camp, Inside this temporary library, we provided toys, games, and reading materials—provided by my American friend and some other individual NASL friends. Earthquake victims were brought to this camp from Sindupalchowk, Dolakha, and Kavre districts and there were a total number of 1,600 families. There were 368 school-aged children. Schools refused to admit them. So, we decided to conduct classes inside this temporary library. With much difficulty, volunteers were teaching these classes. Some children did not want to come to the class, because they were constantly beaten by their parents. Every day, screaming and crying could be heard inside the camp. When we went to see them, parents left them in the camp and went to local market for drinking alcohol. They were always drunk from morning till evening. They lost all sense of control.

We then went to visit them early in the morning. For many days, we held meetings with these children including other parents. We danced with them, we shared stories with them, we sang and gave sociopsychology classes, classes on child psychology and child's rights classes and encouraged them to study hard, so that one day they will become great politicians, medical doctors, lawyers, and so on—in order to make them feel that they will have a bright future, and have someone to love and care for them. If you do not take good care of these children at this stage of their life, they will suffer a lot as they grow older, and their lives could be easily ruined.

Based on my observation, despite many of these children came from poor rural families, there was not so much difference in terms of their intellectual and learning capabilities, when compared with other children from big cities or children raised by rich families.

What we did was to try comforting them by telling them that their misfortunes were indeed caused by a major natural disaster. No one could prevent it from happening. We just have to face the reality together with our combined strengths. We then further convinced them if we continued to provide them with good education, love and care, they could do a lot in terms of protecting themselves when such natural disasters should happen again by using their knowledge—as a way to tell them knowledge and continued learning are most important, instead of using alcohol as an escape from problems. All the above were done in a clean and peaceful environment.

A few weeks later, they all returned to the temporary library with smiles on their faces. Children and parents were dancing, singing, and helping one another out. There were also other people from the local villages, who had been previously denied to education and other learning opportunities, they also came and benefited from our series of educational programs and support services provided via our temporary library.

Two months later, some educated parents also served as teachers and librarians for our temporary library inside the camp. The most valuable thing we gained from the experience is that human beings deserve respect, and everyone is motivated to learn if the right opportunities are provided to them at the right time. Undoubtedly, school libraries and librarians both play an important role in offering such knowledge to both children and parents, regardless how difficult the situations have been.

Materials preparations for school librarians in training

Teacher librarian doing Reading Habit Promotion activities at Kshitiz High School, Kathmandu

Danish teacher librarians and children's book writers are observing children's creativity at Kshitiz High School

CHAPTER 13

HELPING STUDENTS IN THE FOOTHILLS OF THE HIMALAYAS

JEREMIAH O'SULLIVAN

School Librarian, Lincoln School, Kathmandu, Nepal

Please provide a self-introduction including your professional and educational backgrounds. What did you study at university? Are you a second-career school librarian?

My name is Jeremiah O'Sullivan, and I am currently a preschool–12th grade LMS (Library Media Specialist) at Lincoln School, which is a small American international school located in Kathmandu, Nepal. My wife is the preschool–12th grade school counselor, and we have a five year-old daughter (Tegan) who attends school here, and a three-year-old son who stays at home with a nanny. I am originally from Portland, Oregon (US), and I graduated from Oregon State University[1] (USA) in 2004 with a Bachelor of Science (BS) in Liberal Studies with an emphasis of Social Studies (which fit perfectly since I majored in preengineering for a year, photography for a year, and finally social studies/history). After studying for a semester in London, England and then taking a year off to travel through Europe, I returned to Portland, Oregon and completed my Masters in Teaching with endorsements in middle school mathematics, and middle/high school social studies, graduating from Concordia University (Portland)[2] in 2006.

From 2006 to 2009, I taught middle school math for two years at West Orient Middle School[3] in Gresham, Oregon and then high school history at Sam Barlow High School[4], also in Gresham. From 2009 to 2012, I taught

---

[1] Oregon State University—Homepage. Available at: http://oregonstate.edu/.

[2] Concordia University (Portland)—Homepage. Available at: http://www.cu-portland.edu/.

[3] West Orient Middle School—Homepage. Available at: http://westorient.gresham.k12.or.us/.

[4] Sam Barlow High School—Homepage. Available at: http://www.sbhs.greshamsd.schoolfusion.us/.

middle/high school social studies and also served as the MUN (Model United Nations) Director for the 400-student MUN conference our school hosted at a small international school in Taichung, Taiwan. (I was director for our school hosted MUN in Taiwan and I am the head advisor for the MUN Travel Team that has done local and international conferences.) From 2012 to present, I have worked at Lincoln School[5] in Kathmandu, Nepal. The first year I taught middle school social studies, but at the same time, I took online courses to complete my Masters in Library Media from Portland State University.[6] I was able to take online courses and also completed the first half of my practicum/student teaching experience with Jennifer Alvey. For the last three years (currently on my fourth), I have been the LMS here.

Are you a second-career librarian—meaning that you have other jobs before entering a career into school librarianship?

I was a middle/high school math and social studies teacher for seven years before becoming a librarian, so while I changed professions, I still consider myself a "teacher" who works at a school. I do not necessarily feel like a second-career librarian, but I guess people would see it as that.

Choosing a career in school librarianship, was it an active choice out of personal interest or it was by chance and circumstance?

Wanting to become a librarian and enrolling in Portland State University's Library Media Master's program was 100 % an active choice out of personal interest for me. However, falling into a librarian job at my current job was chance/fortune because the previous librarian was in her last year at our school. I was not aware she was leaving when I started my program.

Are you currently working as a solo librarian in the whole school?

I currently work as the preschool–12th grade librarian for the entire school, which has 240 students and I have one clerk who helps me.

---

[5] Lincoln School (Kathmandu, Nepal)—Homepage. Available at: https://www.lsnepal.com/.
[6] Portland State University—Homepage. Available at: https://www.pdx.edu/.

Could you describe the social backgrounds of your students? What do a majority of their parents do for work?

Of the 240 students, about 120 are in elementary school and 120 in secondary. Roughly 25% are American, 25% local Nepali, and the other 50% come from 30 to 35 different countries all around the world, but very few are from South America. Many of their parents are diplomat/embassy, NGO, or foreign business working in Nepal. The local parents have a variety of jobs, but many of them own large companies in Nepal.

What is the economy of Kathmandu mostly based on?

It is based mostly on textiles and tourism.

What is the ratio between male and female school librarians working in your part of the world?

In my PSU (Portland, or USA) library classes, I would say it was one male to five females. At the International Association of School Librarianship (IASL) library conference I attended in Bali (which included both IASL members and local librarians) it was closer to one male to 20 females. Our school is part of SAISA (South Asian International School Association) and of the ten schools, and I believe it is about one male to five females. I am not sure about local libraries here.

Could you describe your typical day at work as a school librarian?

The typical day involves having library time with one or two elementary classes, inputting new physical or Kindle books into our catalog, helping co-teach a middle school (MS) exploratory class (currently working on having students create Google Cardboard virtual reality (VR) and integrating it into their classes), keeping both school-wide subscriptions and student information in the catalog up-to-date, sorting/weeding/organizing both the elementary and secondary library, creating in-library resource lists, finding online resource lists for teachers, giving book talks to secondary classes, and trying to help teachers as much as possible.

Do you need to take up any classroom teaching duties, in addition to fulfilling your roles as a school librarian?

In my first year as a librarian, I had to teach one section of 8th grade Social Studies. This was difficult because, even though I try to follow the flexible librarian approach, I still had elementary teachers that wanted fixed 45-minute library times, but the class I taught was on a rotating schedule. Also, I was spending at least one block per day planning for my social studies class and then another block teaching the class. Currently, the my classroom teaching duties are voluntary ones that have to do with subjects that pertain to the library or technology, and I try to co-teach those so I can leave when needed.

When planning lessons and classes with other teachers, how much do you collaborate and work together?

While there is quite a bit of bouncing ideas off each other in person or on a shared online document, in the end, the teacher ends up doing a majority of the actual planning. Teachers at this school are both fortunate that they all have a teaching assistant, but it is also unfortunate that there is only one section of each subject (at all grade levels). Recently, I have been supplying them with online/physical resources more than sitting and lesson planning with them. (This is not to say that I never help lesson plan, but just on average.)

As a school librarian in your region, is there a nationwide or region-wide syllabus or curriculum that you need to follow, in terms of performing your work as a school librarian? If not, do you think it is feasible to implement a region-wide syllabus for school librarians? The absence of such a syllabus—do you think it is an advantage or disadvantage?

We do not follow a set syllabus or curriculum in the library, but we do have standards that we use. A couple of years ago, our school began implementing the ACS Technology skills and some of the International Society for Technology in Education (ISTE)[7] standards. These are in Rubicon Atlas and are used when teaching library lessons. I am

---

[7] International Society for Technology in Education—Homepage. Available at: http://www.iste.org/.

currently in the process of introducing American Association of School Librarians (AASL)'s standards for the 21st century learner to teachers and administrators in hopes of having them to replace our current ones. I feel like the AASL standards are broad and adaptive enough that they could easily be implemented for the region, but I would hesitate to implement a solid syllabus since all the international schools in our "region" are so different.

What are the expectations amongst your students, other classroom teachers and the senior management in the school library, and in you—in the context of supporting the overall learning and teaching, as well as the development of other recreational activities of the whole school?

I am not told to specifically do anything, but I know that as an active preschool–12th grade librarian in the school, I am expected to be on a variety of committees. I try to do as many as possible while also leaving some time to be flexible and get in classrooms or prepare lessons or resources for teachers.

Please give a list of successful library programs/events initiated by you as a school librarian. Where do you get your ideas and inspiration from when planning library events/research projects?

Yearly book week, multiple author visits per year, board game club, MUN club, book club, and so forth

What are some school library projects that you are currently undertaking?

I recently created a Literacy Week committee that is meeting to help plan a full week of literacy events. I am also in the process of planning a trip to the Netherlands with a dozen students who will take part in MUN. Also, I have taken it upon myself (along with the help of the elementary principal) to catalog every single book in all the elementary classrooms. Additionally, I am working with our IT director to create a school-wide Digital Citizenship Curriculum.

What are the major challenges and difficulties faced by you as a school librarian?

One of the biggest challenges for me is being the only school librarian at the school and the only American international school librarian in the country. I would love to meet (in person) monthly with other librarians, but this is not possible. Because of this, I have tried to create a group called the Kathmandu International School District (KISD), which is a group of librarians from the American school (mine), the British school, the Christian international school, and more recently one of the local schools. Unfortunately, many of the other schools are either underfunded or run by librarians who have been trained (or not trained) on the job, and the meetings have ended up being two or three of the other librarians asking for help and advice on what to do with their library or how to best serve their population. While I have not minded helping as much as possible, it is not exactly what I expected KISD to become and we have been meeting about once a semester instead of once a month.

What are some aspects of your job as a school librarian that you find most rewarding?

I find it most rewarding when I can make a passionate reader of out of a student who used to be a reluctant reader. I also find it incredibly rewarding when I receive compliments from teachers about how proactive I am in getting new and helpful resources into their hands. Multiple teachers have commented on this and it is nice to be appreciated at such a small school.

The professional knowledge, skills, roles, and other job-related competencies for a school librarian—have they undergone major changes in your region in the last five to ten years? In your opinion, what is the future for school librarians in your region?

Since I am only starting my fourth year as a librarian, this answer is a little tricky, but I see the librarian's role expanding into the classroom more and more by helping teacher integrate appropriate technology into their classes; as more and more teachers rely on online subscriptions and technology to organize.

Are there any library organizations that you are part of and if so, does that involvement benefit your work? Do you work closely with the town library?

I am a member of American Library Association (ALA)[8]/AASL[9] and have been a member of IASL[10], but need to renew that one. The most beneficial group that I am a part of is the LM_NET listserv from syr.edu. I do not work at all with the town library since it is all in Nepali.

Do you think having a passion for librarian work is something that can be considered a calling or it is something that could be developed through experience and exposure?

I definitely think that having a passion for librarian work is internal for some people, but at the same time, those people need to develop this passion into something that will benefit those around them. A passion for research, reading, and technology does not benefit others unless you know how to teach, instruct, or facilitate, and bring out the best in people. I do not think I would be as good of a librarian if I had not taught in the classroom for seven years prior to becoming a librarian.

What kind of attributes does a motivated and successful school librarian always possess?

It is important to be flexible, understanding, and proactive. It is also important to ask questions, become involved, and to continue to find current best practices.

As a school librarian, do you sometimes feel that you could choose to work very hard or do nothing at all because in the end, you would still get paid the same amount of salary?

Of course. But, it is like that with lots of professions, teaching included. I could choose to sit around all day and stay in the library, letting teachers come to me, but most of my enjoyment comes from getting back in the classrooms or working closely with students/teachers to make their job/learning/educating as fun and easy as possible.

---

[8] ALA (American Library Association)—Homepage. Available at: http://www.ala.org/.

[9] AASL (American Association of School Librarians)—part of ALA—Homepage. Available at: http://www.ala.org/aasl/.

[10] IASL (International Association of School Librarianship)—Homepage. Available at: http://www.iasl-online.org/.

Throughout your career as a school librarian, did you ever have any regrets or second thoughts?

So far, after three full years, I have no regrets other than I did not become a librarian earlier. I think that teaching for three or four years, I could have become a librarian, but I am glad of the route that I chose.

If the school decided to lay off the school librarian or to close down the school library completely, what kind of impact do you think it would have on the overall learning and recreational needs of the whole school community?

As a whole, the community would lose out on a valuable resource that encourages lifelong reading for pleasure, helps teachers in the classroom, and advocates for the best resources for both students and teachers. Overall, it would be removing a central hub that both serves as an educational hub that promotes pleasure and learning at the same time.

School libraries/school librarians and inquiry-based learning, do you think they go always hand in hand?

I think they have the opportunity to always go hand in hand, but it does not mean they always do. Having taught inquiry-based learning in both math and social studies, a librarian's role can be extremely important since they can help teachers create and plan the enquiries, making it easier on everyone.

What are the core skills and knowledge for a school librarian that would never become outdated, regardless how advanced technologies become?

My biggest skill/knowledge that I have been focusing on is not about finding results, but finding the right results, so I guess it would be "research skills." Anyone can do a random search on any platform and get results back, but that does not always mean people are going to get the right results. Students/patrons need to be able to conduct research with confidence and have a rough idea of what they are looking for. A lot of this has to do with the correct search terms, correct advance search features, and getting reliable sources returned. In the past, librarians were there to help you find your information when search results were at a minimum (librarians used to have to find outside sources for librarians) and today, it is the librarian's job to help you find the correct information when search results are at a maximum.

Regular classroom teacher versus and school librarian in your region, which one do you think has a more optimistic and promising career path and career progression?

I feel like I am spoiled being at a small international school because of the excitement that both students and teachers bring to school each day. However, it will always be easier to cut the specialists, no matter how imperative their position is to student success because you are really only cutting one position. Every student *needs* a teacher, but people always think they can do without a librarian because of the lack of immediate and evidential impact that a librarian can have. Igniting a passion for reading, helping teachers plan for lessons or units, or ordering appropriate books or online materials can easily be overlooked since the librarian is often the only one involved in those things.

Having served as a school librarian for such a long time, did you ever have any second thoughts or regrets?

I have only served as a school librarian for four years (out of my 11 years in education), and I have no regrets about moving from the classroom to the library—mainly because I am still in the classroom almost every day, but just in a different role. Sometimes, I wish that I would have become a librarian earlier, but then I think back to my years in the classroom and feel like it was more beneficial, overall, to have been a classroom teacher in multiple subjects and grades than to become a librarian earlier on.

Given that you had a second chance, is there anything you would have done differently?

No, I feel like the path I chose was one that was a best fit for me, and I would not have done anything different. I loved teaching in the classroom and feel that because I taught in the classroom for seven years, it has given me a better understanding as to what teachers want and need.

Jeremiah O'Sullivan
School Librarian, Lincoln School,
Kathmandu, Nepal
*With his wife, Allison,
at a Nepali festival*

Elementary school library

Secondary fiction/hangout area

A TEACHER-LIBRARIAN WITH A MISSION OF CREATING INFORMATION LITERATE GLOBAL CITIZENS!

ZAKIR HOSSAIN

EUROPEAN International School (EIS),[1] Ho Chi Minh City, Vietnam

Please provide a brief self-introduction by telling us about your professional and educational backgrounds. What did you study at university? Are you a second-career school librarian? Did you have other careers before becoming a school librarian?

My name is Zakir Hossain, and I am a Teacher-Librarian, Researcher, and an advocate of library-based lifelong learning. I earned my Bachelor of Arts in Political Science, Philosophy and Islamic History and Culture, and also a Master's degree in Information Science and Library Management from the University of Dhaka[2] (first class-first position). Currently, I am working at European International School (EIS) Ho Chi Minh City, a Nobel Education Network[3] school, as a teacher-librarian and Diploma Program Extended Essay Coordinator. Prior to EIS, I worked as a Manager of Library and Information Center at the Southeast Asian Ministers of Education Organization (SEAMEO) Regional Training Center[4] in Ho Chi Minh City, and Singapore International School, Vietnam. I also taught in primary

---

[1] EUROPEAN International School (EIS)—Homepage. Available at: http://www.eishcmc.com/.

[2] University of Dhaka—Homepage. Available at: http://www.du.ac.bd/.

[3] Nobel Education Network—Homepage. Available at: http://www.nobelalgarve.com/nobel/nobel-education/.

[4] Southeast Asian Ministers of Education Organization (SEAMEO) Regional Training Center—Homepage. Available at: http://www.vnseameo.org/.

schools in my own country, Bangladesh, before doing my Postgraduate studies. I am a lover of the Internet, travelling, and realistic fiction books—a geek at heart!

I started my career as a primary school teacher (science and computer), before obtaining my Master's degree. Two years ago, I worked as a full-time English as a Second Language (ESL) and Information and Communication Technologies (ICT) teacher at Singapore International School. So, a few ups and downs had occurred.

Choosing a career in school librarianship, was it an active choice out of personal interest? Or it was by chance and circumstance?

Well, after the completion of my undergraduate degree, by chance, I chose to study Library and Information Science (LIS) without knowing anything about this discipline and the career opportunities. However, once I started the program, I was dreaming of myself in the position of a university librarian or working as a faculty in a library school. Therefore, I can say I was quite ready to be a librarian. Although I did not think about working in the school library arena, but now, I love that—especially in international school librarianship.

In your country, is it mandatory for every single public or private school to be equipped with a school library? In addition, is it mandatory for all school libraries to be managed by a professionally qualified school librarian? Or there are cases that a school library is only managed by a regular classroom teacher, who is overseeing the school library as some kind of extracurricular activity?

I am originally from Bangladesh, but currently working in Vietnam. Neither Bangladesh nor Vietnam has a well-equipped school library and professional teacher-librarian or a school librarian. According to the governments of both countries, there must be well-equipped school libraries which should be run by professional librarians, but the reality is just opposite except a few exceptions.

I am one of the contributors for International Association of School Librarians (IASL) World on the Webpage for Bangladesh and Vietnam. The following information has been taken from the IASL Window of the World (WOW) document (https://www.iasl-online.org/WOW):

Bangladesh
Most primary schools have no libraries; some secondary schools have library facilities, with teachers serving as librarians. The school library collections range in size from 500 to 5,000 books but most libraries do not have any budget (World Encyclopedia of Library and Information Services by Robert Wedgeworth). Schools in the larger cities have books, but those in the small villages are lucky to have even a few volumes. Fewer still have libraries. When there are books, most are outdated. The Commission on National Education has emphasized the importance of libraries in education, but the government has given little funding for school libraries (Hossain, unpublished). Where school libraries do exist, in secondary schools, they have generally been neglected, and the total amount of money allotted for the library was always small. In almost all schools no provision was made either for the construction or for the expansion of the library building and, where there was such a provision, it was altogether meager, with the result that school libraries all over Bangladesh face an acute accommodation problem.

Vietnam
For 2014–2015 school year, Vietnam had 27,541 Primary, combined primary and lower secondary and secondary schools (north 14,227; central 3,227; south 9,997) of which 24,746 have libraries (north 12,927; central 2,839; south 8,980). There are about 26,578 library professionals and paraprofessionals (north 13,807; central 3,047; south 9,724) are working in the school library sector where most of them do not have library and information science (LIS) degree (Teachers serving as librarians). The annual budget for school libraries is still poor, for example, in 2014–2015 school year the total amount of budget was 202.8 billion VND (Vietnamese dong, about 1USD = 23000 VND). Average budget allocation per school is about six to nine million VND depending on school location and students. Overall, the situation of school libraries is not satisfactory as only 13,000 (out of 24,686) meet national standards. Only about 43 % of the 26,000 school library custodians in Vietnam are professional librarians (Nguyen, 2015; Hossain, 2016).[5]

Overall, the school library situation in Vietnam is not satisfactory as only 13,000 (out of 24,686) meet national standards. Only about 43 % of

---

[5] Nguyen, Thi Thu Phuong. (2015). Vietnam Country Report on School Library Development. PreConsul Workshop on School Library Development in ASEAN Country. Bangkok.

the 26,000 school library custodians are professional librarians (adapted from Hossain, 2016). Most of the schools allow their students to go to their school libraries just once a week on average because of a shortage of librarians and the small size of libraries (Hossain, 2016).[6]

Are you currently working as a solo librarian in the whole school?

Yes, I am the only teacher-librarian for the whole school consisting of 400 students and 80 staff. It is a K-12 school. I have an assistant who helps me mostly, in circulation and cataloging, and her background is a Bachelor of English. The Homepage of our school library EIS Learning Resource Centre is available at: https://eishcmclrc.wordpress.com/.

Are a majority of the local public schools in Vietnam also equipped with a school library that is managed by a qualified school librarian?

A good number of public schools have libraries mostly without any qualified librarians. The literacy rate is quite high in Vietnam and in my area (Ho Chi Minh City), it is highest in the country, that is, 97.9%. I do not know if there are laws that punish the parents for not sending their children to schools in Vietnam. The statistics (Nguyen, Thi Thu Phuong, 2015)[7] shown in table below might help explain the school library situation in Vietnam.

Region	Number of schools	Number of school libraries	Percentage (%)
North Vietnam	14,272	12,927	90.6
Center Vietnam	3272	2839	54.6
South Vietnam	9997	8980	50.4
Total	**27,541**	**24,746**	**49.3**

Could you describe the social background of your students? What do a majority of their parents do for work?

---

[6] Hossain, Z. (2016). Towards a Lifelong Learning Society Through Reading Promotion: Opportunities and Challenges for Libraries and Community Learning Centres in Vietnam. *International Review of Education*, *62*(2), 205–219. https://doi.org/10.1007/s11159-016-9552-y.

As an international school, students of 30 different nationalities, come from various socioeconomic backgrounds. Although the vast majority are local Vietnamese, there are lots of Korean and European students in the school. Most of the parents are business people and some of the others are high officials.

Could you describe your typical day at work as a school librarian?

Since I am a teacher-librarian, I have to perform two-folded responsibilities as a library manager and information literacy (IL) specialist. I schedule daily classes with primary classes and once a month with the middle years' students. Every week, I also have one lesson with Diploma Program students. Besides, I have to cover classes when a particular teacher is absent. I also meet with students who were caught for plagiarism or academic dishonesty and explain them how to avoid plagiarism and so on. I do co-teaching with classroom teachers—especially on International Baccalaureate (IB) units of inquiry. Finally, I update the library website, answer e-mails, and make sure that the library resources are displayed appropriately.

Do you need to take up any classroom teaching duties, in addition to fulfilling your roles as a school librarian?

Yes, I do. I teach IL, with a particular focus on Internet search techniques, academic honesty, citation, plagiarism, academic writing, research skills, how to use library catalog/databases, and so forth. I am also the extended essay (an independent 4,000-word research project mandatory for IB diploma students) coordinator, and therefore have to spend a huge amount of time with students as a teaching staff to make sure everything is going well with students research projects. I also frequently offer sessions for faculty on academic honesty, role of teachers, how to use Turnitin, Web 2.0 tools for teaching and learning, how to use library resources, citation, and so forth

As a school librarian in your region, is there a nationwide or region-wide syllabus or curriculum that you need to follow, in terms of performing your work as a school librarian? If not, do you think it is feasible to

implement a region-wide syllabus for school librarians? The absence of such a syllabus—do you think it is an advantage or disadvantage?

No, to my best knowledge there is no specific curriculum/syllabus for school librarians–at least in Asia-Pacific region. Well, to have a standard curriculum might be helpful particularly for the new school librarians that will guide them through. However, the challenge is to keep this syllabus current, as "key knowledge and skills" yesterday may not be so today or tomorrow for students or for school librarians. I personally believe such syllabus will give librarians huge advantages regardless of their years of experience. Honestly, I do not see any disadvantage of such an enterprise. If librarians want, they can go beyond that syllabus. Nevertheless, I think it would also be wise to think about curriculum syllabus design for school librarians, such as a syllabus for IB or Cambridge schools librarians.

What are the expectations amongst your students, other classroom teachers, the senior management in the school library, and in you—in the context of supporting the overall learning and teaching, as well as the development of other recreational activities of the whole school?

The common expectation is to make the library a vibrant learning space where students and the wider community feel welcome and safe. Another common expectation is ensuring the access to resources as conveniently as possible through printed and online platforms. As an IB continuum school, my management team is concerned about the IB requirements, and whether the library plays a central role not only for learning but also for other social gathering or recreation.

Please give a list of successful school library programs (supporting students' overall learning and teaching of other teaching staff) initiated by you as a school librarian, and why do you think they are so successful and well received by the school community as a whole?

I could give you a list of examples of successful school library programs, and they are as follows:

Reading aloud—younger students love that, as they just listen while the classroom teachers or school librarians reading to them. This helps improve students' listening skills as well. Students are also encouraged to compete by answering the questions based on the stories or books being read to them;

Creating your own shelf-markers—these kinds of activities help students think that the school library is an integral part of their school life. When they come to the school library and use their shelf-markers, they feel some kind of ownership over the school library and its collections. More importantly, it is a fun activity for the students to be engaged in and they simply love it;

Book display—very important I would say, because it is effective for promoting literacy as our students find seeing all the books clumped together visually overwhelming! Like browsing the Internet, when users come to the school library they first browse around the school library, and would usually first select the ones, which have been out on display in most cases!

School-wide citation continuum and how to teach citation right from the Grade 1 and Academic Honesty, in general, plagiarism and academic writing, in particular—academic honesty is one of the core components of the IB Curriculum, therefore teaching young students how to appreciate and respect the work of others by acknowledging their intellectual debts through proper identification of sources. The language and terms used for teaching the students may differ, but the core aims are to teach them how to appreciate and respect academic works carried out by others;

Web 2.0 tools for teaching and learning—a lot of teachers are not technologically-savvy, and in this case the school librarians can introduce to the other classroom teachers how to master different educational technologies, which could help improve the educational experience of the students, as well as making their lessons and teaching more effective and efficient.

Research skills—how to narrow down a topic—these skills are the prerequisite of Middle-Year Program (MYP), personal project, and DP extended essay, therefore classroom teachers and the school's senior management could see how school librarians are involved serving as teaching partners—helping students identify a research/assignment, as well as locating relevant materials, to be used as references for writing their assignments and other external tests.

What are the major challenges and difficulties faced by you as a school librarian?

Time and enormous demands of the community are the major challenges. Since I am the only teacher-librarian in the school, I really cannot focus on a particular age group; rather, I have to work with all the age groups. There is a lack of guidelines, mentorship and training opportunities as I am a solo person who has to drive several programs by myself!

Which parts of your job as a school librarian did you find most rewarding?

Reading aloud, searching skills. I also enjoy teaching research skills.

The professional knowledge, skills, roles, and other job-related competencies for a school librarian—have they undergone major changes in your region in the last five to ten years? In your opinion, what is the future for school librarians in your region?

There is an extensive body of literature discussing the evolving new roles amongst school librarians/teacher-librarians being information literacy experts toward technology integration leaders and digital citizenship and data literacy coaches. School librarians are aware of these changes and their new roles as leaders in the new learning and fast-evolving Internet environment. My understanding is that there is a huge change in the field of librarianship due to the technological revolution. The challenge is, however, to cope with the required professional knowledge and skills in the digital age as what were "key knowledge and skills" yesterday may not be so today or tomorrow. In my region, I foresee the school librarian as a future technology integrator and digital driver along with their previous role as IL facilitator especially in international schools.

Having a passion for school library work, do you think it is something that is inborn (some people would say it a calling) or it is something that could be developed over experience and exposure?

To me it is something that is developed over time through experience. However, you still need to devote a fair amount of time to current in this profession.

What kind of attributes does a motivated and successful school librarian always possess?

Lifelong learning and caring attributes.

As a school librarian, do you sometimes feel that you could choose to work very hard or do nothing at all—at the end, you would still get paid the same amount of salary? People are sometimes promoted because of their seniority (only they have been here longer), and not because of how well they do their jobs?

I have never had this sort of feeling. I consider myself as an active person, and therefore, I am moderately successful in my profession from a teacher to university librarian to teacher-librarian and now as an Extended Essay Coordinator. I have always been rewarded for the works that I have done.

What are the core roles and responsibilities of an Extended Essay Coordinator?

Based on the IB Requirements, an Extended Essay Coordinator is expected to carry out the following:

- Ensure that students can determine the subject for their extended essay from the approved extended essay subjects before choosing the topic;
- Ensure that each student has an appropriately qualified supervisor, who is a teacher within the school;
- Provide supervisors and students with the general and subject-specific information, and guidelines for the extended essay;
- Provide supervisors with recent extended essay subject reports;
- Ensure that supervisors are familiar with the IBO (International Baccalaureate Organization)[8] document Academic honesty;
- Explain to students the importance of the extended essay in the overall context of the Diploma Program;
- Explain to students that they will be expected to spend approximately 40 hours on their extended essay;
- Set internal deadlines for the stages of producing the extended essay, including provision for a concluding interview (viva voce).

---

[8] International Baccalaureate Organisation—Homepage. Available at: http://www.ibo.org/.

- Ensure that students have been taught the necessary research skills and been provided appropriate training for supervisors.

Other roles are as follows:

- Prepare, publish, and distribute Academy guidelines for the Extended Essay to both teachers and students, these being inclusive of:
  - Overview of the Extended Essay task,
  - The allocation of supervisors or team of supervisors,
  - Timelines,
  - Roles of students and staff.
- Distribute text "Extended Essay" to each student;
- Conduct information sessions for students and parents as required;
- Monitor deadlines;
- Monitor ManageBac and documentation of staff–student interaction;
- Provide guidance for students and staff regarding the research process, academic writing, and operation matters;
- Provide workshops at appropriate points in the research process or as needed by individuals;
- Supervise submission of the Extended Essay;
- Supervise storage of the Extended Essay task in the school library.

Throughout your career as a school librarian, did you ever have any regrets or second thoughts?

As a librarian, I did, but as a teacher-librarian I do not. I love this profession as I can perform my role as a teacher and a librarian.

If they were to lay off the school librarian or to close down the school library completely, what kind of impact do you think it would have on the overall learning and recreational needs of the whole school community?

I do not see any possibility of this kind of initiatives as the school library is expected to serve as the hub of learning and inquiry. As a dynamic organization, the layout and services of school libraries have changed and will be changing over time because of community demands and technology.

School libraries/school librarians and inquiry-based learning, do you think they go always hand in hand? In the school environment, true inquiry-based learning could not be carried out without a proper school library that is managed by a professionally trained school librarian?

Not always, of course. It is almost impossible to create an inquiry-based learning atmosphere without an experienced and trained school librarian. In school librarianship, teaching ability is more important than the library management; therefore school librarian should possess some sort of teaching qualification.

Regular classroom teacher versus and school librarian in your region, which one do you think would have a more optimistic and promising career path and career progression?

Generally, regular classroom teachers, but obviously, it varies case by case. For public schools, the situation is worse and has less possibility to improve. But in international schools, they have a good future. For example, I am very optimistic in my career as a teacher-librarian.

Are there any other interesting stories that you would like to share with the readers?

To survive as a teacher-librarian or school librarian, one should consider oneself as a lifelong learner, be proactive, and practice evidence-based teaching. I have participated in lots of online courses and webinars to keep myself updated. Following social media groups and participating job-alike are other ways to keep oneself current. Community advocacy would be another area all librarians should focus on. Be caring and enjoy sharing what you have known. You cannot get all the teachers on board, but target one or two, and they will be your aid.

Further Reading:
Zakir Hossain (personal homepage) – Available at: www.theresearchtl.net

Zakir Hossain
European International School (EIS), Ho Chi Minh City, Vietnam
At the SEAMEO RETRAC Library and Information Centre (2013)

Popular series at the European International
School HCMC in 2016-17 AY

Book display at European International
School HCMC in Summer 2016

PART III

Australia

CHAPTER 15

LIBRARIES—A FLOURISHING GARDEN OF LIFE

ANDREW DOWNIE

Teacher-Librarian, Fairfield High School,[1] Fairfield, New South Wales, Australia

Please provide a self-introduction by telling us about your professional and educational backgrounds. For example, are you a second-career school librarian—meaning that did you have other careers before becoming a school librarian?

I did majors in Modern American and European History. I did school library methods as part of my Graduate Diploma in Education. Later on I gained a Postgraduate Diploma in Teacher Librarianship. My initial employment as a teacher-librarian in 1978 was due to my Graduate Diploma in Education in Library method and the Graduate Diploma has taken me to another level.

I began working as a teacher-librarian at a Catholic Systemic High school in Sydney from January 1978 to May 1984. I then took on a full-time teaching role from May 1984 until the end of 1985, again in a Catholic High School in Sydney. From 1985 until July 2005, I worked in various roles in different private enterprises. And from July 2005 to the present, I have been employed as a teacher-librarian at my current school in South West Sydney. My current school is a government high school and it has the distinction of being the most multicultural school in New South Wales.

My various roles have included Chairperson, Oliver User Group International Student Coordinator; Practicum Coordinator/HOPP; School Photo/ID Card Coordinator; Homework Centre Coordinator; Member of the School Finance, Welfare, W.H.S. & Technology Committees; and School ARCO Officer.

---

[1] Fairfield High School—Homepage. Available at: http://www.fairfield-h.schools.nsw.edu.au/.

From 1985 till July 2005, you worked in different private enterprises—does it mean that you resigned from being an educator, and you spent the following 20 years working for a commercial company? Could you tell me why you chose to undertake such a major career change? And why you finally decided to venture back into the school librarianship in 2005?

At the end of 1985, I needed a break from teaching. At one school I worked at, I did not think my contribution was being valued. So, I wanted a break from teaching. I also had a political bug in me so over the next 20 years—I took on various different roles including working as a Research Assistant for Federal Politicians. I later worked in educational sales companies. So, in one respect, I did not really leave education. I just broadened my horizons. I looked at education from a variety of standpoints. When I decided to return to school librarianship, I did so knowing that I had a great deal more to offer to our staff, and students.

After spending 20 years in different private enterprises, and you decided to re-enter school librarianship as a career—did you have to undertake major cuts in terms of salary, fringe benefits, and pension plans?

The only major financial loss for me was in terms of future superannuation entitlements. But, as I plan to keep working for as long as I am physically able to, hopefully this will not be a major issue.

Your current school being the most multicultural school in New South Wales—could you describe the profile of your student population? What are the major challenges in terms of working for a multicultural school? Anything that you need to do differently as a school librarian, when you are serving a large number of culturally and ethnically diverse students?

My current school, according to 2015 figures, has students born in 64 different countries, speaking over 70 languages. So it presents a great challenge for both me and our teachers. In any year, over 50 % of our students are refugees, mainly from the Middle East—Iran, Iraq, and Syria. We also take many refugees from Asia—Vietnam, Thailand, and from Africa. We also have a significant Filipino population in our school although they are not refugees.

In terms of the library, you purchase a lot of what could be described as ESL/Primary level resources because whilst our students may speak three, four, or five languages, their level of English compared to students in private schools would be judged as poor. However, we still have a handful of students who excel, by any measure, in the Higher School Certificate (HSC) each year. In fact, when I am asked to describe my proudest achievements at my school, one story I always mention is the fact that in the 2010 HSC, two of our students who were refugees from Iraq, from a single parent family, came first and second in the state, in the same two subjects in the same HSC year. Truly an amazing achievement.

International Student Coordinator—please describe in details your main roles and responsibilities at your School?

International students are, if you like, our fee-paying students. We have about 25 international students who pay AUD$ 14,000 each for each year of their tuition. The State Government gets about 75% of that and our school gets the other 25% to help cover student expenses. My role as International Student Coordinator is to act as their "Roll-call Teacher," to make sure they pay their fees on time and their visas, etc., are up-to-date and that they comply with all Government regulations covering international students, particularly, their attendance at school and their academic progress at school.

Most of our international students come from Asia, for example, China, Vietnam, Thailand, and so forth

The students who came to Australia as refugees, compared with the students who grew up in Australia—how are their reading needs, and learning motivations and needs different from each other?

Obviously, the students who came to Australia (as refugees) have far greater needs than the students who grew up in Australia—their needs are more intense. In particular, their literacy levels in English are so much lower. So, as a school librarian I have to make sure that when I purchase library books, these books need to reflect this need. For example, we bought many elementary readers, picture books, and graphic novels. Often, because they come from a poorer country—economically—their motivation to learn is much greater, because they realize that they have an opportunity to learn that their older siblings or parents did not have. In addition, there are a number of support teachers employed to assist other teachers in this regard as well.

As a school librarian, how do you cope with a group of students who have such diversely different backgrounds, literacy skills, and learning needs?

In a nutshell, with great difficulty! You do the best you can. You will never satisfy all their needs, but with the support of other teachers and purchasing the appropriate resources, you can go a long way in terms of helping them. I also have to be conscious of the fact that with our students from our Intensive English Centre, most of whom are recent arrivals in Australia, these students prefer to read books, which is what they are used to rather than introducing them straight away to digital resources. That comes later, once they have a better understanding of English. Telling real life stories to them helps as well. In my own case, it also helps that my wife is a refugee from Vietnam in 1979. So I, can show empathy towards our refugee students. I also have a love of history and I use that love when I can. For example, one of my Filipino library monitors who attended our school from 2009 to 2014, I had a special teacher/pupil relationship with. We often shared stories with each other about one of our shared heroes: General Douglas Macarthur. On the other hand, some of our Vietnamese students come from what I will term "Devout Catholic" families. Given that I am Catholic, I often talk about religion with these same students.

Could you name three to five titles from your library collection printed, digital, or audiovisual (AV) resources, that are very popular and highly circulated amongst your students of refugee background?

1. An abridged version of *Gone with the Wind*—I am thinking of a two-volume set of this novel, which has about 70 pages each and is A5 in size. I have many classic novels in this format.
2. I have purchased a large number of graphic quarto fiction books, which are very popular, for example, *Snow White and the Seven Dwarfs.*
3. Cut-down versions of any Roald Dahl novel, that is in simple language and A5 in size.

In other words, any good novel in simple language and preferably with some graphics.

What is the minimal professional qualification for working as an entry-level school library in Australia?

Now it is a Graduate Diploma in Teacher Librarianship, and to be eligible for membership of the Australian School Library Association.[2] This qualification is a combined degree of five years full-time.

In Australia, during the time when you were still in elementary/secondary school, was each school also equipped with their own school library?

I completed my schooling between 1960 in kindergarten and 1973, when I did my year 12 twice. I did not score enough marks to get into university from my first attempt in 1972.

During my primary school days, 1960 to 1966, no school that I am aware of had a school library. In high school, as in primary school I went to Catholic schools, and we got our first school library when I was in year eight (that was 1968).

School libraries first became popular, between 1972 and 1975 when the Whitlam Labour Government was in Power Federally.

"School libraries first became popular, between 1972 and 1975"—could you tell me since when (around what year) the public schools in Australia began to recognize the importance that all school libraries should be managed by a qualified and professionally trained school/teacher-librarian?

I guess from 1974 ff., after the federal government issued a report into school libraries; it was known as the Horton report.

The emergence of school libraries and school librarians in Australia—was it meant to respond to the changing face of the educational needs amongst the public schools in Australia—that is the inquiry-based learning processes are gradually replacing the traditional education system, which emphasized passive learning, rote memorization, and spoon-feeding of knowledge with teachers as the sole providers—is my understanding correct?

I would say to a large extent, yes. But there was also the recognition of the value of reading in a child's education. The value of reading in a child's education has been highlighted in many studies, both here in Australia and elsewhere over the years.

Are you currently working as a solo librarian in the whole school?

---

[2] Australian School Library Association—Homepage. Available at: http://www.asla.org, say, au

Yes, I am the sole "Teacher-Librarian" in my current school.

As a school librarian in your region, is there a nationwide or region-wide syllabus or curriculum that you need to follow, in terms of performing your work as a school librarian? If not, do you think it is feasible to implement a region-wide syllabus for school librarians? The absence of such a syllabus—do you think it is an advantage or disadvantage?

In Australia, from 2017 we will have a National Curriculum from kindergarten until year 10. Though, in my home state of New South Wales we have our own version of a National Curriculum. So, as a teacher-librarian, I work to support the teachers who have the responsibility of teaching this National curriculum. So, YES, it can work! On the other hand, I think that a teacher-librarian should be able to vary what they teach, depending on the needs of their individual students.

What are the expectations amongst your students, other classroom teachers and the senior management in the school library, and in you—in the context of supporting the overall learning and teaching, as well as the development of other recreational activities of the whole school?

I think the staff and students want me to support the teaching and learning that goes on at school in any way I can, and to make the library a friendly and welcoming place for staff and students. Having totally remodeled our school library in 2016, that is, new furniture, shelving, carpet, and repainting, we intend, in 2017, to look at other services that the school library can offer, above and beyond what it already does.

School libraries/school librarians and inquiry-based learning, do you think they go always hand in hand? In the school environment, true inquiry-based learning could not be carried out without a proper school library that is managed by a professionally trained school librarian?

The short answer is yes! All the research shows that school libraries function much better when staffed by an appropriate level of a number of professionally qualified teacher-librarians.

Please give a list of successful library programs (supporting students' overall learning and teaching of other teaching staff) initiated by you as a school librarian?

In 2015, for example, I implemented a library skills program for our students in our Support Unit. The students in the Support Unit are students who are intellectually challenged and in some cases they suffer from varying degrees of Autism. The program from both the teachers and students point of view has been labeled a success so that it will continue in 2017 and beyond.

What are the major challenges and difficulties faced by you as a school librarian?

I suppose one always has the challenge of trying to be all things to all people, but sometimes you cannot for reasons such as time and budgetary considerations. In late 2015, my school principal asked me to develop an Action Plan to take the school library into the 21st century. Stage one of that process was completed by the end of 2016, and in 2017 we will look at what further services the library can offer to support the teaching and learning at our school.

Budgetary constraints—do you think the other faculties such as art, music, PE, IT/computer science teachers also have the same kind of budgetary constraints as you do as the School Librarian?

Budgetary constraints are always an issue in every school. However, given that in most schools, the principal has the final say on budgets, his/her attitude goes a long way in determining budgetary constraints in most schools. For example, I have never known a school principal who is, say, trained to starve the science faculty of funds. But, I know of many schools where a principal decides to cut a library budget by 50% or more. This is particularly prevalent in primary schools. In my own case, my current school principal has installed an active finance committee, who has the final say on budgets. And I got myself elected to the finance committee so that I can best represent the needs of the school library. It also helps me that my current school principal was a library monitor when he was at school, so I do not have to convince him of the value of a school library.

In 2016, the school library had an increase of 25% in its school budget. With a new principal, our first official increase in ten years. For 2017, I had another 10% increase in the school Library budget. But we have had a massive increase over the past six months in the number of refugee students enrolling in our school particularly from Syria. So the demands on the library will be greater. If I need more money, I can apply via the finance committee. However, my previous principal was very generous with the library as well. In his time many things that we bought for the library, for example, Blinds/iPads, etc. were bought out of other funds. So, it depends somewhat on how the school principal runs the school finances.

What are the male/female ratios amongst school librarians in Australia? The school librarian profession is predominately female—do you think it has to do with the job nature (school librarianship) or the unpromising career path (limited career progression) that is discouraging many men to choose a career in school librarianship? For example, if a promotion opportunity should come up, it is very likely that the senior management would choose to promote a subject leader (say, Mathematics, English, or Science) who is leading a group of junior subject teachers to prepare students for state-wide examinations, instead of giving this promotion to the school librarian?

You have basically answered your own question. Whilst some teacher-librarians have gained promotion opportunities including a former Director of the Education Department in NSW. She was once a teacher-librarian for a short time—we are mostly overlooked.

This is because our role is misunderstood, and because we do not teach full-time, we are seen in many schools as not being "real teachers"—that is, little or limited marking, No reports to write and no six periods a day. This, I believe is because, we, as a profession, have not marketed ourselves properly. Note my e-mail signature.[3] I have had this signature for about two years now, and it is truly amazing, how teacher attitudes at my school have changed towards my role in the school.

Which parts of your job as a school librarian did you find most rewarding?

---

[3] Andrew Downie's signature:

Teacher - Librarian
Fairfield High School
Chairperson Oliver User Group
International Student Co-ordinator
Practicum Co-ordinator/HOPP
School Photo/ID Card Co-ordinator
Homework Centre Co-ordinator
Member of the Multi Cultural Day Organising Committee
Member of the School Finance, Welfare,
W.H.S. & Technology Committees.
School ARCO Officer

The friendships that you make with individual students as you help them meet their academic goals and the fact that many of those friendships last well beyond the time when the student has left school. I love the fact that some students who have left school continue to visit me at school, and we discuss their progress at university amongst other things.

The professional knowledge, skills, roles, and other job-related competencies for a school librarian—have they undergone major changes in your region in the last five to ten years? In your opinion, what is the future for school librarians in your region?

In New South Wales, Government school libraries are implementing a new library system in 2015–2017. We are going from a DOS-based system to a web-based system. This new system allows us to offer staff and students so many more services so that the sky is truly the limit in terms of school libraries and teacher-librarians in New South Wales.

According to your experience, if the school is facing financial difficulties, the school library is always the first victim of budget cut—is that common?

No! The first victim of any school cuts depends on the view of the school principal and his/her executive and/or finance committee. Certainly, school libraries have been cut when the principal does not value the contribution of the school library to the life of the school. And, this is where the teacher-librarian comes in. I believe that one of my roles is to make sure that the school principal so values the school library that the school library is the last place that the principal thinks of cutting.

How would you go about to make sure that the school principal would value the school library in context of the whole school community?

There is no short answer. That requires a whole essay. I could share with you various antidote stories about how one wins over their school principal. Obviously, each school library is different in terms of the role it plays in the whole school community. Teacher-librarians are up against it, if the school principal does not value the importance of reading, and believes that Google is the panacea for all students research needs. And sadly, some principals think that way.

However, to ensure that the school principal values the school library in terms of the whole school community, I believe that one has to consider the role of the library and the role of the teacher-librarian in the life of the school. When considering the role of the school library, or for that matter, the role of any organization, I believe that it helps if one looks at the past. And when I think about that, I am reminded of some very wise words of one of my heroes, Pope John XXIII.

On January 25th, 1959, when he was announcing his decision to a group of stunned Cardinals in the Vatican, that he had decided to call an Ecumenical Council, which became known as Vatican 2, he said and I quote: "We are not here to guard a Museum, but to cultivate a flourishing Garden of Life."

Just as those words were relevant to the Catholic Church in 1959 as they are today, so they are relevant to school libraries today and yesterday. In the past, too many teacher-librarians saw their role in running a school library as "Guarding a museum." To give but two examples.

1. I did my undergraduate degree at Macquarie University[4] between 1974 and 1977. And in my fourth year—1977—I had to spend 1 day per week in a school library. One day, I arrived at school at about 8:15 a.m., and the teacher-librarian told me that she was so glad that I had arrived early. She asked me to tidy the shelves for the next 2½ hours as the school principal was coming into the library at 11:00 a.m. to talk to her. She said to me: "What the hell does he want; he never comes to the library." And she was in a panic. So, I did as I was asked, and right on 11:00 a.m., the principal came into the library, I was introduced to him and he stayed for about 15 minutes. Throughout the whole year, that was the only time I saw him except once on school assembly. This librarian guarded the school library like a museum and when I drove home that afternoon, I said to myself that if I ever got a job in a school library then the school principal would be welcomed at any time whether the shelves were tidy or not. If the shelves were not tidy, then at least he/she knew that the library was being used.

2. About 30 years later, I was at an in-service course, and I met a school principal for the first time. He asked me how things were,

---

[4] Macquarie University—Homepage. Available at: https://www.mq.edu.au/

and I told him that I was about to do my first stock take at my current school. He laughed. I asked him what was wrong. He said that when he was a deputy principal at one school, the teacher-librarian asked him if she could close the library for three weeks so that she could do stock take. He told me that, so as to avoid an argument, he said yes. He told me that this particular teacher-librarian did not welcome anyone into the library, that all the books were in perfect order, and that he thought that the stock take should take only two days. Even if the stock take took six days, what an attitude for a deputy principal to have towards the school library and the teacher-librarian. Obviously, the teacher-librarian did not market the school library in any way shape or form.

I believe that the school library should be: "a flourishing Garden of Life." So, not only does it house relevant books, computers, iPads, and so forth, but also that the school library should be multifunctional, and teachers and students need to realize that the school library is the place where it all happens. Over and above that, the teacher-librarians need to market themselves. Consider the following examples of how I market myself.

1. Ever since the beginning of last year, when I redesigned my e-mail signature, no teacher ever accuses me of having the easiest job in the school as they did in the past. When they come and see me to ask for something, they usually begin by saying, "I know you are really busy, but have you got a moment for this?"
2. Since 2014, I have e-mailed all the universities in New South Wales the number of student teachers that the school will take each year. Another one of my jobs to organize. Having done that, I then e-mail the principal and the three deputies stating what I have done, the numbers involved and that no university had a right to complain, as they had done in previous years the number of students we have taken.
3. A third way I promote the school Library to my Principal, members of the school Executive and staff is through the use of Twitter. Often, a tweet with a photo attached is a far more effective in letting the powers at be know what is going on in the library on any given day than trying to organize a meeting with them and telling them what went on in the Library on that day.

With these few examples, you think my work is not valued and the school library is not valued. We are forever having special events in the library. It has all to do with marketing without being seen as sucking up to your school principal. If you get my drift.

If they were to lay off the school librarian or to close down the school library completely, what kind of impact do you think it would have on the overall learning and recreational needs of the whole school community?

In one word, devastating! School libraries these days are so much more than about books. In my own case, I was once asked by somebody at a library conference, who was just starting her library qualifications, to describe my role. I described that I see my role in this order: (A) counselor, (B) teacher, and (C) librarian—that means not just to students but staff as well.

If a young man is inspired to enter a career in school librarianship (after finishing his MLIS) and asks you for your career advice, what would you say to this young person? Anything he should watch out for?

I would advise him to follow his passion, that is, he has to be passionate about the role school libraries can play in students' education. I would also sell the benefits of being a teacher-librarian. For example, effectively, you are your own "boss." And I would stress the unique relationships a teacher-librarian can build with individual students. I would share some of my own experiences with him.

In your opinion—having a passion for school library work—do you think it is something that is inborn (some people would say it a calling) or it is something that could be developed over experience and exposure?

The short answer is both. But you need to have a work ethic, which most people do not have in my opinion. In my own case, it comes from both my parents who, when they were alive, instilled in me and my seven brothers and sisters, a passionate belief in the pursuit of excellence in terms of everything that you do. Always do your best and always aim for the best that you can afford.

Throughout your long career as a school librarian, did you ever have any regrets or second thoughts?

The short answer is "No"! Though, obviously, there have been lots of times throughout my career when, on an individual day, I have said to myself: "I could have done that better," for example, dealing with students and staff or teaching a particular library lesson, and so forth.

Do you have any strategic plans worked out for developing your school library in the next few years?

Stage 2 of the Library update—that is improving the technological output of the school Library is almost complete. It will be completed in the next two weeks just in time for Christmas and the 2018 school year. It includes a new Data Projector with a much larger screen, two 75-inch TV screens which teachers will be able to use so that they can display wirelessly to their students Powerpoint and images from their devices to their students when they are teaching a lesson in the Library.

We have also purchased 18 laptops, which will be used in one of the back rooms of the Library—that is, like a computer lab. And finally, we are in the process of adding a Makerspace upstairs in the Library. In fact, the first equipment for the Makerspace was delivered yesterday (1st December 2017).

In 2018, I will develop a plan for stage 3 development of the school Library. All exciting stuff.

REFERENCE

Library Learning Resource Centre, Fairfield High School—Homepage. http://www.fairfield-h.schools.nsw.edu.au/curriculum-activities/library-learning-resource-centre.

Andrew Downie
Teacher-Librarian, Fairfield High
School,1 Fairfield, New South Wales,
Australia

Eight permanently fixed iPads in the school library adjacent to the computer section.

One view of the non-fiction area outside the teacher-librarians office.

CHAPTER 16

MAKING A DIFFERENCE TO STUDENT LEARNING THROUGH SCHOOL LIBRARIES

MADELEINE JANE VINER

Head of Library Resource Services, Kilvington Grammar School, Ormond, Victoria, Australia[1]

Please provide a self-introduction and tell us about your professional and educational backgrounds. What did you study at university? Are you a second-career school librarian—meaning that did you have other careers before becoming a school librarian?

My name is Madeleine Jane Viner, and I am known professionally as Jane Viner. I have been a teacher-librarian from the beginning—I was trained as a secondary teacher in Library and Geography at Melbourne State College—Teachers' College (does not exist today)—from 1975 to 1979. I graduated as a Bachelor of Education (Librarianship) with education teaching qualifications to teach in secondary schools both Geography Year 7 to 12 and Library Year 7 to 12 as a teacher-librarian. I was one of the first teacher-librarians and part of a government initiative to fill the new Commonwealth (Federal) Government libraries that were built as a stand-alone state-of-the-art, late 1970s and 1980s brown brick buildings in the middle or on the edge of the school to cater for students reading and research. The Federal Government built these buildings without a thought about how to staff them, hence the new course component of "teacher librarianship," which was available to me from 1976 with me swapping from Science—Biology and Geography to libraries, as this seemed a fabulous new option, which was not there in 1974. I was able to keep the

---

[1] Kilvington Grammar School—Homepage. Available at: http://kilvington.vic.edu.au/enrol/contact-us/.

Geography—my favorite subject and which I am still teaching today in 2016.

A large number of teachers came into the colleges from schools and upgraded their Diplomas into Bachelor's degrees, and many of these became teacher-librarians. Sadly, some of these were classroom escapees, so they did not always give Principals of schools, their staff, and students the best option or enthusiasm for students—my cohort chose library for its benefits, not as an escape as we had not even begun our teaching careers.

In 1994 to 1998, I undertook a Master's of Information Management and Systems at Monash University[2] on a scholarship and graduated the following April in 1999. This degree did not give me any more credit in my school, despite it enabling me to also teach Computer Science as it was called then, but the teacher at my school would not relinquish her classes so I looked elsewhere and successfully applied and was appointed as Director of MLC Libraries in March 2000.

I have taught full-time as a teacher-librarian, Geography teacher, and home-group teacher since July 1980, and I had one year maternity leave in 1985. I have held library management and leadership roles since 1984 in schools throughout the Melbourne Metropolitan area.

- 2014–2016 ongoing—Head of Library Resource Services—Kilvington Grammar School, Ormond
- 2000–2013—Director of MLC Libraries—Methodist Ladies' College, Kew
- 1994–1999—Faculty Head of Information Resources—St. Leonard's College, Brighton East
- 1993—Curriculum Resources Coordinator—Sandringham Secondary College, Sandringham
- 1986–1992—Head of Library—Mordialloc-Chelsea Secondary College, Mordialloc
- 1984—Acting Head of Library—McKinnon Secondary College, McKinnon
- 1981–1983—Teacher-librarian—Ballam Park Technical School, Frankston North
- 1980—Teacher-librarian—Aspendale Technical School, Aspendale

---

[2] Monash University—Homepage. Available at: http://www.monash.edu/.

I have also been awarded the Australian School Library Association Teacher Librarian of the Year Award 2017.

Choosing a career in school librarianship, was it an active choice out of personal interest? Or it was by chance and circumstance?

As mentioned above, it became a personal choice but I "fell into the course" due to a desire to change from Biology to another subject in my second year and the option of teacher librarianship was available. I had a part-time, casual job at the local public library, so I had some prior knowledge, and I was recommended by the course advisor to apply—the rest is history! I love every day at my new school as I am back at the grass roots level—working with students rather than managing a big team of staff, and being removed from the day to day and the coal face.

The school that you are currently working or, is it a private school or local government school?

Private independent school—Church school affiliated with the Baptist religion but it is open to students and staff from all or no religion.

Are you currently working as a solo librarian in the whole school?

Yes—I am the only teacher-librarian for 800 students (as of 2017)—from kindergarten to grade 12 (ages 3-17/18) and staff.
    As a secondary teacher-librarian, I am continually learning about working with primary-age and kindergarten-age children.

What kinds of social backgrounds do your students come from? What do a majority of their parents do for work?

We have limited information about the specific work done by parents. But based on Government records required to be completed each year, the majority of our parents work in Category 1 jobs—professional, senior management, government admin, and defense. We do also have a sizeable number of parents that fall into Category 3 jobs—tradesmen/women, clerks and skilled office, sales and service staff—this includes many families running family businesses especially in the building sector.

Could you describe your typical day at work as a school librarian?

- Teacher-librarian known as TL.
- Support for individual students, class groups, and classes.
- Support for teachers and support staff.
- In person, online, and via e-mail.
- Class preparation for research or reading classes. For example, introduction to research, resources, and bibliographies for VCE Senior History.
- There is a lot of class preparation for introduction to an author or genre for year 7 to 10 reading classes.
- Suggested books for class sets, for example: primary and kindergarten classes at Kilvington are now supported by the classroom teacher not me except for special occasions, for example, Book Week.

Library opens 8:00 a.m.	Opened by library technician	Library technician on duty*
8:15–8:20 a.m.	Arrive at school for roll call	Library technician on duty*
8:30–8:40 a.m.	Roll call for tutor group (home group)	Year 7, 8, and 9–20 students
8:40–9:40 a.m. Session 1	Library administration	Library technician on duty* TL in office
9:40–10:40 a.m. Session 2	Library duty—includes class support, shelving, loans, research, and reading queries	ELC—year 12 Individual students Booked classes
10:40–11:05 a.m. Recess	Library duty	Student free time (50–80 students year 4–12)
11:05–12:00a.m. Session 3A	Early lunch/recess break	Library technician on duty*
12:00–12:40 p.m. Session 3B	Teaching—tutor/house/school assembly	Library—classes with classroom teacher—no TL support
12:40–1:30 p.m. Lunchtime	Library duty	Supervision, students free time Year 7–12
1:30–2:00 p.m. junior school lunchtime	Library duty	Supervision, activities, e.g., Book Club on a Thursday, year 2–6

Library opens 8:00 a.m.	Opened by library technician	Library technician on duty*
1:30–2:35p.m. Session 4	Library duty	Year 7–12 classes, individuals support for year 7–10 English classes—reading 1 h per fortnight
		Support for Subject classes—e.g., VCE 20th century history research intro
2:35–3:35 p.m. Session 5	Library duty or private study or year 9 Geography teaching	Year 7–12 classes, individuals (as above)
After school till 5:00 p.m.	Meetings	Library technician on duty*

TL: teacher-librarian
* In 2017 this after-school duty is carried out by a teacher or teacher librarian.

Do you need to take up any classroom teaching duties, in addition to fulfilling your roles as a school librarian?

Yes—the following all reduce my time as a TL part of the role of being a TL is to be recognized by the school as a teacher so these responsibilities are expected in some schools; otherwise they could employ a librarian for a lower salary and more working hours.

- Year 7, 8, 9 Tutor Group Teacher—roll call daily plus 4 × 40-minute lessons per fortnight.
- Year 9 Geography—3 × 60-minute lessons per fortnight.
- Year 10, 11, 12 Sport Coach—4 × 60-minute lessons per fortnight plus bus travel 2 × 50-minute per fortnight.

As a school librarian in your region—is there a nationwide or region-wide syllabus or curriculum that you need to follow, in terms of performing your work as a school librarian? If not, do you think it is feasible to implement a region-wide syllabus for school librarians? The absence of such a syllabus—do you think it is an advantage or disadvantage?

No syllabus, but in Victoria and in Australia we have the Australian Curriculum, and this needs to be supported by the TL by finding relevant resources for classes and their teachers.

Primary—a syllabus may be an advantage if it is taught as part of subjects rather than a stand-alone separate curriculum where the knowledge is not often transferred.

What are the expectations amongst your students, other classroom teachers and the senior management in the school library and in you—in the context of supporting the overall learning and teaching, as well as the development of other recreational activities of the whole school?

The expectations are high, and I have increased these expectations by offering an inclusive library program and place to all—"the library lounge room." As a small team of three, we are very proud to be part of the Australian Great School Libraries Honor list—see our website (just released in April with our new library information management system) http://library.kilvington.vic.edu.au.

Please give a list of successful library programs (supporting students' overall learning and teaching of other teaching staff) initiated by you as a school librarian?

KILVINGTON GRAMMAR SCHOOL

- Kilvington Reads Festival May 2016 (for ELC, Junior and Senior Students, Parents) Inaugural event,
- Promotion at Senior School Assembly May 16—Kilvington Reads, Premier's Reading Challenge for ELC—Year 10, Access-it,
- Kilvington Library Webpage—ideas and promotion of digital and print resources, bibliography template, Premier's Reading Challenge—via school e-mail, newsletter, class introductions,
- Access-it—promotion and demonstration to all teaching and support staff at staff briefing,
- Access-it—library information management system April 2016—implemented and promoted to whole school community,
- Access-it promotional bookmark—designed by library technician,
- Access-it enabled student reserves,
- Access-it enabled student and staff reviews,
- Junior School Library Leader 2016,
- Junior School Library Book Club 2016—Year 6 students meet weekly,

- Great School Libraries Honours list 2015—nominated by students, staff and parents,
- Kilvington student video—for EduTech 2015—written by students, produced by library technician,
- Presentation at EduTech 2015 on a panel on curriculum curation and individual presenter on "Collaboration in school libraries" Brisbane June 2015,
- Kilvington Library Webpage—ideas and promotion of digital and print resources, bibliography template, Premier's Reading Challenge,
- Introduction to library digital resources—from arrival in Feb 2014 and ongoing,
- Bibliography introductions to Year 7–12 classes as requested,
- Research introduction to Year 7–12 classes on a variety of topics from VCE History to Biomes,
- Reading class introduction on specific genres, authors, holiday reading,
- Premier's Reading Challenge for Year 7–10 2016,
- Continue with PRC for ELC—Year 6,
- Labeling of library collection with PRC labels to improve access,
- Weeding of fiction and nonfiction collection by library team,
- Continual improvement of collection with increased budget including student and staff suggestions,
- Book Week 2015, 2014—special program for students,
- Ref—Publication of an article about collaborating with classroom teachers—in school library journal,
- Viner, Jane "2013 onwards: Moving schools, changing roles, rekindling the spark of collaboration and teaching to make a difference to student learning outcomes" FYI Spring 2014.

METHODIST LADIES' COLLEGE

- Wiki development by library team,
- Introduction and implementation of Lib Guides,
- Presentation at IB, International, national and local library conferences,
- Publication of journal articles promoting library programs,
- Viner, Jane "Teacher-librarians—an essential species to connect, integrate and lead curriculum change in our digital world" ASLA XX111 conference. The curriculum experience: connect, integrate,

leader Sept 28–Oct 1 2013 Hobart, Tasmania. http://www.slide-share.net/ASLAonline/lead-curriculum-change   Accessed   Sept 2015.

What are the major challenges and difficulties faced by you as a school librarian?

TIME! LACK OF TIME! Conflict of interest with multiple hats—sometimes too many hats reduce the effectiveness of the main hat—library.

Which parts of your job as a school librarian did you find most rewarding?

Working with students, finding them a gem to read, turning on the reading enthusiasm, finding them a resource to help their study, demonstrating the value of a bibliography, showing how wonderful Ebsco and Encyclopedia Britannica can be, and the variety and quality of their resources, etc.

Being asked back to a class or to another class to introduce their students to resources, etc.

Positive feedback from students and staff—tonight I received an e-mail from a student thanking me for helping her with our online resources—she was delighted it was working from home.

Having my expertise valued by highly experienced teachers.

The professional knowledge, skills, roles, and other job-related competencies for a school librarian—have they undergone major changes in your region in the last five to ten years? In your opinion, what is the future for school librarians in your region?

School librarians in Australia have reduced in primary schools dramatically and many government and some independent secondary schools have replaced their TLs with support staff. It often comes down to the individual school and their Principal and the value adding they see their TL giving. During the 2000s, I was the Chair of the IB Australasian Library Group and the situation was healthy in the IB Independent and International schools but I do not know the situation in 2016.

Having a passion for school library work, do you think it is something that is inborn (some people would say it a calling) or it is something that could be developed over experience and exposure?

It has become my calling, and I feel I really excel at this as I am very enthusiastic in the role and can see that I am value adding in each interaction each day. With the right mentor, placement, and support this could be developed in a teacher that was willing to give it their best and take advice.

School libraries/school librarians and inquiry-based learning, do you think they go always hand in hand? In the school environment, true inquiry-based learning could not be carried out without a proper school library that is managed by a professionally trained school librarian?

Ideally, the TL will be involved with inquiry-based learning, and a professionally-trained TL is employed to work with classroom teachers and students to improve student-learning outcomes.

If they were to lay off the school librarian or to close down the school library completely, what kind of impact do you think it would have on the overall learning and recreational needs of the whole school community?

In our school, I hope this would never occur as the library is now held in high regard by the school community. It is essential that we continue to offer supportive and educational programs that make a difference to student learning outcomes. We need to educate our school communities of the value of a TL and professional library support staff.

As a school librarian, you could choose to work very hard or do nothing at all–at the end, you would still get paid the same amount of salary—do you agree?

I do not agree as in schools today we are all accountable—the more time release from direct timetabled classes, the more one needs to prove the worth of the role—this is a luxury not a given—TLs should not just expect this time release but value and treasure it and prove their worth to their Principal and school community.

FURTHER READING

Kilvington Grammar School Library—Homepage. Available at: http://library.kilvington.vic.edu.au/#!dashboard.

Madeleine Jane Viner
Head of Library Resource Services,
Kilvington Grammar School, Ormond,
Victoria, Australia1

Junior school students reading and relaxing in junior library

Senior Fiction—for older students and staff—McKie Library

CHAPTER 17

THE LIBRARY WORKS AT BEING THE HUB OF THE SCHOOL

KEVIN WHITNEY

Library Manager, Kew High School, Victoria, Australia

Please provide a self-introduction by telling us about your professional and educational backgrounds. What did you study at university? Are you a second-career school librarian? Did you have other careers before becoming a school librarian?

I completed a Bachelor's of Arts degree at University of Melbourne[1], majoring in English and Politics, along with two years of Classical Studies and one year of Philosophy. I have had a number of part-time jobs in high school, university including in hospitality as a waiter and barman, as a proofreader for University of Melbourne publications and in the freight department of a Melbourne fashion house.

After university, I initially worked for the Australian Bureau of Customs (managing the office for investigation officers) and as a personnel clerk for a major telecommunications company.

I, then, gained a Diploma of Education at Melbourne State College, and taught for four years at two Melbourne schools. I had always wished to pursue a career in the performing arts so returned to studies at drama school, and then worked as a singer/actor for 20 years (with professional opera companies, television drama, theatre roles, concert performances, advertising campaigns, etc.) I also had extensive experience in corporate event management work.

During that time I worked as a relief and contract teacher in 25 schools. I worked in all subject areas, and had a range of short contract jobs. Working with teachers in all subject areas allowed me to understand the different needs staff and students have and how the library can flexibly meet them.

---

[1] University of Melbourne—Homepage. Available at: http://www.unimelb.edu.au/.

I completed a Masters of Education (Teacher Librarianship) degree through Charles Sturt University[2] in 2005.

Choosing a career in school librarianship, was it an active choice out of personal interest? Or it was by chance and circumstance?

I have friends who are trained librarians and also worked extensively in schools as a relief librarian. I quickly realized that I liked the work and wanted to pursue it full time.

Are you currently working as a solo librarian in the whole school?

I am the head librarian at a state high school with 1,200 students. I have library support staff but I am currently the only trained accredited librarian.

Could you describe your typical day at work as a school librarian?

Whilst there are consistent routines to follow, each day can be varied. The first library staff member arrives and opens the door (so anytime between 7.30 and 8:00 a.m.) We have three trolleys of laptops available for students, and so we have to see if the batteries have charged overnight. I have to check e-mails to see if any changes to the day affect us (staff absences, excursions, etc.), and reply to requests for equipment, books, and library bookings. We loan laptops to student teachers and relief teachers, and answer any technical and resource questions that may have. I contribute to all ongoing library work when needed—cataloging, book covering, management of library book-ings, maintenance of the online video library, including recording programs, and promoting fiction to students. We attempt to deal with requests when staff appears in need of assistance. I regularly visit staff rooms to drop off requested resources as well as catching up with all staff for professional and other conversations. We always encourage staff to ask us for assistance. The school library hosts year level assemblies, theatre company perfor-mances, visiting speakers, staff meetings, senior examinations, and so forth. To streamline these events we need to relocate timetabled classes to other rooms, rearrange furniture, welcome visitors, and provide them with neces-sary equipment. We supervise the large number of students at lunchtimes.

---

[2] Charles Sturt University—Homepage. Available at: http://www.csu.edu.au/.

And, of course, every day, we focus on encouraging reading. The library is busy at lunchtime with students completing class work, reading, playing games, watching programs on their laptops and generally socializing.

Do you need to take up any classroom teaching duties, in addition to fulfilling your roles as a school librarian?

I teach extras classes when staff is away. I usually relocate the classes to the library. I also do yard duty outside the library in the school grounds.

I am the coordinator of the annual school awards night, which is held at Melbourne Town Hall in the Melbourne[3] central business district (CBD). This involves a range of organizational activities such as liaising with the venue staff, booking buses and equipment, organizing payment of invoices, convening meetings, coordinating rehearsals, and working with the stage manager on the evening.

I am the staff representative for occupational health and safety (OHS) for the school, working with the Assistant Principal to meet the safety goals of the school (which involves completing forms, repairing faulty equipment and facilities, and attending OHS conferences to maintain my professional knowledge).

I am also the returning officer for staff elections within school. Staff nominate to be part of interview panels for new teachers and committees such as the Local Administrative Council which oversees school activities.

I am involved in the school swimming and athletics sports every year as well.

As a school librarian in your region, is there a nationwide or region-wide syllabus or curriculum that you need to follow, in terms of performing your work as a school librarian? If not, do you think it is feasible to implement a region-wide syllabus for school librarians? The absence of such a syllabus—do you think it is an advantage or disadvantage?

We are incrementally implementing an Australian Curriculum in every subject of the curriculum throughout the high school years. The library's role is to be aware of the changing resource needs as well as the role

---

[3] Melbourne Town Hall—Homepage. Available at: http://www.epicure.com.au/venues/venue/melbourne-town-hall.

of technology across the school. We work with teachers in each subject area to support them as the curriculum changes of each subject areas are confirmed.

What are the expectations among your students, other classroom teachers, the senior management in the school library, and in you—in the context of supporting the overall learning and teaching, as well as the development of other recreational activities of the whole school?

The school administration renovated the original library space with the view of establishing the library as the hub of the school. They also have given me the freedom to administer the library on my own terms. Influenced by my experiences with libraries, and working as a classroom teacher in my subject areas as well as a relief teacher in all other areas, I realized that every subject area needs different kinds of support from the library. We are very proactive in offering services and resources. We regularly search through our collections to provide staff with books that they may be unaware of. Then, of course, we encourage staff to request our assistance as well as source new resources. We have a set of flexible procedures that allow us, as is reasonably possible, to meet school needs as efficiently as possible. We also encourage students outside class time to use our games collection and gather in the library to work and socialize. We host school debating competitions and show televised events and programs on the large library screen. The Olympic Games and World Cup soccer are very popular and well attended.

Please give a list of successful library programs (supporting students' overall learning and teaching of other teaching staff) initiated by you as a school librarian?

I have maintained the existing library programs that were here when I arrived as manager eleven years ago. One change has been to focus on the purchase of novels suggested by the staff and students. This has meant that there are fewer books that we buy, but they are rarely borrowed. One innovation we brought in is Technology Free Friday. We ask the students not to bring in or turn on any electronic devices every Friday. The rationale is to encourage some mental and habit space between themselves and their use of technology.

What are the major challenges and difficulties faced by you as a school librarian?

Unfortunately, the status of our profession has been downgraded by state education departments. When I began teaching, every school had several trained teacher-librarians whereas now school principals are employing librarians without teaching qualifications, and even library technicians to staff school libraries. I am extremely lucky to work in a school that values our contributions. I think a challenge we face is promoting non-fiction printed books to staff who have themselves grown up with computers— we are constantly encouraging them to book their classes in to use the excellent print resources we have for class assignments.

Which parts of your job as a school librarian do you find most rewarding?

The flexibility of the individual daily tasks and interactions with all staff and students are highly enjoyable. I love connecting students with litera-ture, and enjoy seeking hard to find resources that staff may have been searching for in vain. I find it rewarding that staff see the library as a happy welcoming place.

The professional knowledge, skills, roles, and other job-related competen-cies for a school librarian—have they undergone major changes in your region in the last five to ten years? In your opinion, what is the future for school librarians in your region?

We are working even more to support the use of technologies in school. I assist the audio-visual library staff expert in loaning, maintaining, and explaining the best use of our digital cameras. We also maintain the 61 laptops that are loaned when students' personal laptops are being repaired. The library space is used in a much more flexible way, and we have to be aware of daily needs as well as advanced bookings (e.g., year-level assem-blies and theatre company performances). We do not create pathfinder sheets as no teachers request them. We now have an online video system, which we are expanding. We have also subscribed to a new online video provider called Click View, so I have attended training sessions. Our online purchases of books have increased. Access to the library staff by phone and e-mail means we are responding to an increasing range of requests in a quicker manner. We also, of course, keep an eye on the Australian Curriculum developments so we can predict changes in textbooks.

As I responded earlier, I am not sure if any government schools will see the value in having a library. I know that privately-funded schools see the importance of having properly trained teacher-librarians as an educational priority. I know there are funding issues, and schools have ever-decreasing sources of funding. But also, there is a shortsighted view that everything is available on the World Wide Web and that books are old technology. Unfortunately, many people in positions of educational administration and decision making are not allowing debate regarding the implications of this thinking.

Your previous careers in the show business and in the advertising company—how have they contributed to your current work as a school librarian?

From my experiences as a library user, my work in other school libraries, my work experience in business (such as bakeries and cafes), and my performing arts work, I understand that people are more enthusiastic and generally positive about an environment that is welcoming and supportive. Working with the public is a kind of performance. You are presenting to the users what is available to them; you are actively engaging them in exchanges and collaborating with them to create possibilities. It can be a very creative process when relationships between library staff and users are established. It does come down to individual personalities—I try to discover the interests of staff and students and talk enthusiastically with them (even if what they are involved in is not in my realm of interests)—it is all about establishing the notion that the library is a place to seek assistance and receive personal attention. Library users are more likely to return when there is a sense of trust and concern for their time and needs—I know, I respond to engaging personalities.

Having a passion for school library work, do you think it is something that is inborn (some people would say it a calling) or it is something that could be developed over experience and exposure?

I was always drawn to great literature, so I was attracted to good libraries with interesting books. I accidentally fell into librarianship, and I soon discovered I was suited to it. Because of the range of skills librarians use, I realized this could be a very satisfying career. I always tell people now that, even though I had wanted passionately to be an actor I really was born to be a librarian. The involving nature of the work, I think, suits people who thrive with these kinds of work activities.

What kind of attributes does a motivated and successful school librarian always possess?

The more I worked in a range of libraries, that more I realized that each librarian had a blend of skills, knowledge, and interests that they uniquely bring to the workplace. Attention to detail and the focus to follow through with the many aspects of library work are important. There can be a demarcation of duties attitude in school libraries—some teacher-librarians may not be involved in library work such as cataloging even when a new resource needs to be processed and delivered to a library user; they may see their role in a purely academic sense. What satisfies me is the search for a resource that someone has been searching without success and finding it—like solving a mystery. The interaction with customers is the cornerstone of our work. I also enjoy regularly reviewing how the library resources are arranged so that they are accessible.

As a school librarian, do you sometimes feel that you could choose to work very hard or do nothing at all—at the end, you would still get paid the same amount of salary? People are sometimes promoted because of their seniority (only they have been here longer), and not because of how well they do their jobs?

I realized ten years ago that school library jobs were in danger. If teacher-librarians did not expand their notions of access to the library and engagement, then they would be replaced by technicians. As Kevin Hennah commented, if businesses ran their operations like many school libraries then they would be out of business. I love being busy and juggling a range of jobs every day. As I told a student this morning the best jobs I had as an actor happened to be the least paid. I enjoy a sense of involvement and growth in work. In our education system today, you are more likely to be employed if you can bring innovation and enthusiasm into the workplace.

Throughout your career as a school librarian, did you ever have any regrets or second thoughts?

I always wanted to be a performer and realized after a number of years that the life it brings did not suit me. I am grateful for the opportunities this career has given me—and continues to give me.

If they were to lay off the school librarian or to close down the school library completely, what kind of impact do you think it would have on the overall learning and recreational needs of the whole school community?

School libraries defend the promotion of literacy. They promote learning communities. They provide havens for people to research information, have social interactions, discover unexpected interests, counterbalance overuse of technology, celebrate the historical and ongoing growth of thought and individual discovery, and so forth.

School libraries/school librarians and inquiry-based learning—do you think they go always hand in hand? In the school environment, true inquiry-based learning could not be carried out without a proper school library that is managed by a professionally trained school librarian?

Teacher-librarians have the educational background to understand the curriculum and what elements of inquiry-based learning are applicable to individual subject areas and activities. The VELS (Victorian Education Learning Standards) outlines the incorporation of elements of information literacy within across the year levels in all subject areas. Teacher-librarians can assist teachers with involving these elements in tasks. I would argue that, whilst it is the domain of the classroom teacher and subject faculties to be aware of these, teacher-librarians play an important role in supporting staff and students with these tasks. As I have already commented, teacher-librarians encourage the use of printed resources as well as websites to seek a range of information sources.

Regular classroom teacher versus and school librarian in your region, which one do you think would have a more optimistic and promising career path and career progression?

Education authorities are unfortunately unsupportive of school library staffing. The increase of leadership positions in schools has drastically decreased available funding and commitment to effectively staffing libraries with trained teacher librarians. As well there is the erroneous notion amongst many administrators that everything useful to research can be sourced from the World Wide Web.

Kevin Whitney
Library Manager, Kew High School,
Victoria, Australia

Exterior of the Kew High School Library

Book display at the Kew High School Library

CHAPTER 18

INVITATIONS ACROSS THRESHOLDS

ROBYN MARKUS-SANDGREN

Library Manager, St Hilda's School Senior Library, Southport, Queensland, Australia

Please provide a self-introduction and tell us your professional and educational backgrounds. What did you study at university? Are you a second-career school librarian? Did you have other careers before becoming a school librarian?

My name is Robyn Markus-Sandgren. I began my career as a Mathematics teacher after completing a Bachelor of Science in Pure Mathematics at UNSW[1] and a Postgraduate Diploma of Education at the University of Sydney.[2] I was then selected to undertake a Diploma of Resource Teaching, with the view to equip me to lead the introduction of innovative programs in high schools. I worked in this capacity for nearly ten years before following a different path, running a small business with my husband, while raising a family and completing a Masters of Educational Studies. My main interests in the Master's program were in developmental psychology and the impact of emotional and social well-being on learning.

After completing my Master's, I resumed teaching Mathematics before being appointed as Library Manager at St Hilda's School[3] on the Gold Coast. For 14 years, I have led the library through technological innovation and the creation of a library space that embraced classroom teaching, research support, wide reading, lifelong learning, and social connectedness. There is at the heart of my vision for a library a sense of participatory responsiveness to change that supports individuals, including the "outliers," as they learn, grow and mature in an evolving community.

---

[1] University of New South Wales—Homepage. Available at: https://www.unsw.edu.au/

[2] University of Sydney—Homepage. Available at: https://sydney.edu.au/

[3] St Hilda's School—Homepage. Available at: http://www.sthildas.qld.edu.au/

I have now retired but continue to work in educational contexts on a contract basis.

Choosing a career in school librarianship—was it an active choice out of personal interest? Or it was by chance and circumstance?

I became a school librarian out of circumstance. I was offered the job at the school to manage a change in the role of library in the school.

Are you currently working as a solo librarian in the whole school?

Our school has 1,250 students from Pre-Preparatory (PP) to year 12. While I manage the databases for the whole school, there is a teacher who plays the role of Teacher-Librarian in the Junior School Library.

Could you describe your typical day at work as a school librarian?

My day typically starts at 7:30 a.m. and ends at 4:00 or 5:00 p.m. from Monday to Friday. I man the Help Desk with a colleague until 1:30 p.m. and do managerial work for the remainder of the day. On the Help desk we assist students and teachers doing research assignments (there are eight to nine classes using the library at any one time), shelve resources, do routine cataloging and accessioning tasks, chase overdues, and support the print and copy station. Behind-the-scene managerial tasks include planning, putting into action and updating of purchasing, budgeting, cataloging and accessioning processes, attending to database updates and integration, interacting with teachers to create and improve library support services, professional reading and writing, and attending to the pedagogical and managerial work associated with library student group (for example, we have two active book clubs) activities. I am also expected to roll mark and sometimes supervise classes without teachers.

As a school librarian in your region, is there a nationwide or region-wide syllabus or curriculum that you need to follow, in terms of performing your work as a school librarian? If not, do you think it is feasible to implement a region-wide syllabus for school librarians? The absence of such a syllabus—do you think it is an advantage or disadvantage?

There is no syllabus at any level to guide or direct our work in school libraries. In our school, students in years PP–6 have scheduled Library lessons, and these are considered part of the English curriculum, which is directed at both the state and national levels. In years 7 to 12, in our school, there are no scheduled library lessons, so the onus is on me to integrate library skills and theory into the teaching and learning as it occurs in our school. There are both advantages and disadvantages to this approach over an externally mandated library syllabus. The advantage is that you can tailor your approach for teachers, for classes, for particular assignments and projects, for year groups, and for individuals. The disadvantages are twofold. First, your library programming appears piecemeal, perhaps invisible and assessing or demonstrating the efficacy of what you do is fraught. Second, your work is multiplied by the number of different approaches that you are able to adopt in the face of the needs and wishes of teaching staff and students.

Most importantly, without a visible syllabus and a mandated role in its development, and relationship with the wider worlds of teaching and learning, your role as a librarian in the school remains entirely a support role with little professional standing.

What are the expectations amongst your students, other classroom teachers and the senior management in the school library, and in you—in the context of supporting the overall learning and teaching, as well as the development of other recreational activities of the whole school?

The library staff in the senior school is considered a part of operations rather than teaching and learning in our school. We have little or no opportunity to lead change in educational practices in the face of professional learning or your own teaching practices, and subject teachers often wonder what you do! They wish to be the experts and see you as servicing their and their student's needs when and only when determined by them. Servicing these needs is not seen as a "profession;" rather, it is seen as a secretarial role and, when it comes to technology, as an additional IT support. It is for this reason that I volunteered to move the library from its own building to the new senior school teaching center, and agreed to be surrounded by classrooms connected directly to the library space itself. Change is happening despite resistance.

While the Library is valued as a safe place for students to go—before and after school (we are open from 7:00 a.m. to 8:30 p.m. Monday to Thursday, 7:00 a.m. to 5:00 p.m. on Friday, and 1:30 p.m. to 4.30 p.m. on Sunday) little, real teaching or learning are perceived to occur in the Library.

I conduct two book clubs and attendant excursions/incursions/speaker and author visits/book fairs. There is no time allocation allowed for these cocurricular activities and teaching staff are not allowed to claim quantum allowance for supporting these activities, unlike Art Clubs, Debating, Sports Clubs, and others. So the teachers do not support any of our initiatives, nor do they encourage their students to do so. Nevertheless, we have busy and active groups of students who do ask for more and more from us and who value, respect, and enjoy the work of the Library.

These facts emphasize the subsidiary role that our administrators see the library playing in the school's teaching, learning, and recreational functions.

Please give a list of successful library programs (supporting students' overall learning and teaching of other teaching staff) initiated by you as a school librarian?

- St Hilda's School Readers' Cup Competition to prepare for the Children's Book Council of Australia Readers' Cup Regional Competition.
- Middle School Book Club,
- The Southport School—St Hilda's Book and Philosophy Club,
- Using Our Databases Library Quiz.

Why do you think these library programs are so successfully and well received by the school community as a whole?

They are well received because they are grounded in the needs, interests, and capabilities of the students. They are adaptive, allow a courageous openness of thought, and provide a way for students to connect to the wider world of others (other schools, libraries, students) who share a love of books, learning, thinking, and most importantly, sharing.

Most importantly, they do not cost the school in terms of staffing I conduct them all voluntarily and in my own time. It is a passion given freely, and shared, that also helps to make such groups/activities successful.

What are the major challenges and difficulties faced by you as a school librarian?

The major challenge is to incorporate library and, in particular, research skills into the curriculum and assessments. It is the place for foundational research skills and the role of online research in senior, further, and life-long learning that struggle to find a place in a crowded curriculum. Both are masked in the consciousness of teachers and students increasingly relying on Google, YouTube, and apps to source information.

Which parts of your job as a school librarian did you find most rewarding?

I enjoy providing access to information and ideas that both amaze and help complete learning tasks. I love sharing books and films with others. I am rewarded, in helping to provide a "third space" for students to be in, in seeing students' love of learning increase, in seeing their interactions with others flower and in facilitating learning, and being democratic, community oriented, cosmopolitan (somewhat multicultural), and self-directed.

The professional knowledge, skills, roles, and other job-related competencies for a school librarian—have they undergone major changes in your region in the last five to ten years? In your opinion, what is the future for school librarians in your region?

Technological change—especially mobile technology—has transformed the ways people interact, socialize, seek information, learn, and express themselves. Naturally, this has impacted on how a school library operates and so what skills a school librarian needs.

In my region, access to books, information, and all online and mobile media is ubiquitous. Yet local expressions of culture are increasingly being drowned out in a kind of global din. Students and their families, schools and their communities need, in my view, to not only be strongly connected to the wider world, but to their own voices and what they see can be said, as part of that wider world. A library should support this approach in the ways in which it connects to different voices—from the different subject areas, to the different age levels, nationalities from which students and teachers come, to families and to community histories. The library is, increasingly, less an object and more an activity, more a space and less a building.

The future for school libraries in my region is for them to strongly connect to ideas and information in whatever ways they can and to support the library staff and spaces that can enable both individuals and educational communities to speak with insight, for themselves. Processes and change should be reflective, being communicated ideally, and felt joyous rather than duty bound. There should be a chance, in library practice, for its responsive element to lead change ahead of (but not without) reflection.

Having a passion for school library work, do you think it is something that is inborn (some people would say it a calling) or it is something that could be developed over experience and exposure?

I think that it is not an either or question. Certainly anecdotally, I have been a curious, questioning, and somewhat interior person from a very early age. However, it was the inquiry-based liberal education I received that enabled me to become an explorer of life and a reflector on life's processes. It is this that provided the basis for engagement with school library work. A library is a place perfectly suited to those who are interested in the world and themselves and the complex inter-weavings that reveal both insights and connections in the whole.

What kind of attributes does a motivated and successful school librarian always possess?

The most important attribute is an interest in and responsiveness to the relationships between things and people. The second most important attribute is a professional understanding of teaching and learning. The third most important aspect is an understanding of developmental psychology, both with respect to individuals and to groups/communities.

As a school librarian, do you sometimes feel that you could choose to work very hard or do nothing at all—at the end, you would still get paid the same amount of salary? People are sometimes promoted because of their seniority (only they have been here longer), and not because of how well they do their jobs?

There are many reasons why some people are promoted rather than others and, as you point out, it may have little to do with how well they do their jobs or how hard they work. This applies in many work places and is not specific to school librarians. As a school library is embedded in a school, it is vulnerable to the workplace culture of the school as a whole. Human relations personnel and teacher perceptions, along with support from the Head of School have as much influence as anything—for better or worse. Few understand what a library is, let alone what it can be, so promotions in school libraries are fraught. The students understand libraries better than most employers.

Throughout your career as a school librarian, did you ever have any regrets or second thoughts?

Only that the teacher–student relationship is special and more difficult to develop across a wide range of students, especially as a manager of a library. I loved teaching and view myself as a teacher and a librarian. Other's perceptions are sometimes difficult to refresh.

If they were to lay off the school librarian or to close down the school library completely, what kind of impact do you think it would have on the overall learning and recreational needs of the whole school community?

The students value the library as a space to meet, relax, and to study in. It is a very important place—not a classroom, not a playground for the students to be, especially if they want to/need to be alone. So, the space itself, even if it only had furniture in it, is an essential place for the students to be and not feel the pressure to be social.

As for the physical resources, many students still love to read printed books and many, surprisingly for some, like to do research from books as well as online. The vast majority of students would miss the library for the absence of free, easy to find, wide reading, and research materials.

Our library's staff members are highly valued as support staff for student's learning and as an adult to chat to, not a teacher and so partisan, but as someone who, nevertheless, knows directly what is being asked of them. The library staff provides support not just with accessing resources, completing citations, and books to read, but also an ear.

Some teachers would miss the support that the library gives to resourcing the curriculum and the research support given to their students as they complete their assignments.

School libraries/school librarians and inquiry-based learning—do you think they go always hand in hand? In the school environment, true inquiry-based learning could not be carried out without a proper school library that is managed by a professionally trained school librarian?

Inquiry-based teaching and learning, and libraries definitely go hand in hand and are essential to each other. To manage a school library that supports inquiry, you need a person who has a history of successful and innovative inquiry-based teaching practice and who has undertaken postgraduate research-based studies—preferably, but not essentially, in library studies. Further library studies are best obtained in response to the technological and pedagogical needs of the school library in its context. Ongoing, professional development is as essential to school librarians as it is to teachers and should be mandated as it is for teachers.

Regular classroom teacher versus and school librarian in your region, which one do you think would have a more optimistic and promising career path and career progression?

Teaching is a more promising career in schools, though there are still possibilities available. However, beyond the school, library professionals, including teacher-librarians, have some added options career wise.

Are there any other interesting stories or experiences of you as a school librarian working in Australia that you would like to share with the readers?

When young people enter a library, especially the first time, they do so as orphans. It is not that they have lost their families or cultures, of course, but that being "at home" is suspended in time and space as they cross the threshold. It is also true that for many teachers, entering the school library is like entering a foreign country, in that it is not their teaching space. The way in which they are met, the space into which they enter, and the possibilities for discovery, learning, making connections, and forming relationships comes to be what the library is. The library, unlike the other spaces for teaching and learning in a school, is ideally, benignly structured, so that the formalities are only there to facilitate access and engagement. At the human scale, a library can be seen to parent being at home in the world (Troy et al., 2014).[4]

Two aspects of the work of St Hilda's School Library that embody these ideas are the formation of the difficult to name Book and Philosophy Club, and the building of the new library itself, integrated into the center of the Senior School teaching area. The inception, growth, and ongoing development of these two projects show the changing nature of the relationships of the library to both individuals and to teaching and learning, to education itself. The library has both been transformed by and has transformed the ways in which education is understood, conducted, and experienced.

The Book and Philosophy Club started out, some 15 years ago, as a book club in which avid students could share what they were reading. They were not reading just the popular books; they were interested in works that challenged them, such as Chuck Palahniuk's Fight Club and Alex Garland's The Beach. They were interested in sharing what they themselves discovered to read and they were interested in sharing their reading with other readers from St Hilda's brother school, The Southport School. It soon became very clear that adolescent boy and girl readers chose to read very different books when left to their own devices. Yet there was the natural interest in sharing. How could the gap between their interests be bridged? When looking at the *ideas* in the books they were reading and the video materials they were watching, we discerned that there were commonalities to be found. We asked the group to look for these ideas and, if possible, to suggest a focus idea per school term and then to pick out of their reading, watching, and learning, this idea, how it was explored and what they thought. When the ideas became the focus, the group broadened, deepened, and became more cohesive. Topics for discussion came out of both school learning and everyday experience and have included nationalism, terrorism, freedom, legacy, sexism, violence, and environmentalism.

We saw that more informal learning, sometimes more learned types of engagement with students could be knitted into the fabric of the more

---

[4] Troy, M.H., Kella, E. & Wahlström, H. (2004). *Making Home: Orphanhood, Kinship and Cultural Memory in Contemporary American Novels.* Manchester: Manchester University Press.

formal curriculum and could complement it and, sometimes expand its reach. It, perhaps, could also transform it.... The idea of making the library an actual part of teaching and learning spaces and opening the spaces up, to become part of the library began to be explored. In the old library, two classrooms, once separate, became fluid parts of the library so that the distinction between what was classroom and library began to be made, opened up, and remade across the day fluidly. Teachers, students, and library staff together, created a living educational, cultural space. The experiment was a resounding success, so when the new senior school building was being planned the library was included as an integral part of it. A whole level of the building became a kind of Third Space, using ideas from Homi Bhabha and Lev Vygotsky, for example, and calling on lessons learnt from our experiment.

The new library is busy from 7:00 a.m. to late in the afternoon, and most evenings. It has an energy and life sustained in connection with teaching and learning practices, and by the amazing work and contributions as individuals of the library staff. The well-resourced library nourishes others through these relationships. Time will tell what the library will become in the future as it is co-constructed by the students, the school staff, and the wider school community. It is, however, the library's commitment to the full expression of humanity, our full exploration of the world, and to honoring the connections between all things that will continue underpin the integrity of the St Hilda's School Senior Library.

Robyn Markus-Sandgren
Library Manager, St. Hilda's School
Senior Library, Southport, Queensland,
Australia

The space to read and think…

The digital and physical collections to inspire wonder…

Teaching and learning at the threshold.

CONCLUSION

BRINGING IT ALL TOGETHER

Different countries have different environments, different stages of development, different economic conditions, different cultural landscapes, and different standards of living, and so on. As their state economies advance, many changes will also happen in the social, economic, and cultural landscapes, as well as in education. Different education policies are developed and enacted in response to the broader social, cultural, economic, and political changes and needs, which are taking place regardless of their pace or population size. Some of these education policies are national, while others are local policies. The different school types are, therefore, decisively shaped by the wider policy environment.

These different educational environments influence school librarianship throughout the world. School librarians with many different levels of education and training, expertise, skill sets, experience levels, and attitudes could be found in a variety of types of schools—from those working in urban schools to those in (sometimes primitive) rural areas—from those specializing in information literacy skills and inquiry-based learning to those devoted to raising the overall literacy level of students in rural areas. School librarians in urban areas often give special attention to raising the reading levels of boys who are not interested in reading; in developing countries, girls who have previously not been able to attend school, may need extra help in improving their reading skills.

Depending on the job nature and responsibilities, as well as the individual school policies, school librarians sometimes carry different job titles, such as, school library (resources) managers, school library consultants, teacher librarians, library media specialists, information literacy specialists, and so on. Recently, in countries such as Australia and perhaps the USA, school libraries are employing Information and Communications Technology (ICT) staff also. Sometimes these ICT staff, even without teacher training, would even replace the trained school librarians. These school library ICT staff are employed because the school library has become so reliant on technology, and often the school librarian has not had enough retraining to cope with all the changes.

As highlighted by Lo and Chiu (2015), "The school library is expected to function far more than just as a quiet place for studying and reading. Meanwhile, the school librarian is expected to serve more than a teacher overseeing the daily operations of the physical library....The changing concepts in school librarianship are closely linked with students' new mode of inquiry-based learning on one side, with new technologies and the Internet environment on the other" (p. 697).[1] Despite the important roles they play as instructional partners, and in information access and distribution, professional practices amongst school librarians outside English-speaking countries have been insufficiently studied. Additionally, few studies of professional practices of school librarians have taken place in languages other than English. There is also a lack of information about school libraries in developing and emerging countries, at local school level. Previous studies were limited to small numbers of empirical quantitative studies on the issue. In fact, the study described in this book is one of the very few qualitative studies, that reports on these kinds of details in depth, and discusses the professional practices and changing roles amongst school librarians on a cross-national, as well as cross-cultural level.

Participation of the school librarians for this interview book project was on a purely voluntary basis. There were non-English-speaking school librarians who wanted to speak out, but, unfortunately, not able to find anyone to help translate their interview answers into English, or their requests to take part in our interview project were denied by the senior management. The limitations of this book lie in the fact that a majority of the school librarians appearing in this book come from international private schools or schools that follow the IB (International Baccalaureate) curriculum. Others come from English-speaking countries, which have a strong school library tradition. In other words, not enough local public schools are represented. The reason for that was because mostly the IB and the international school librarians were willing to speak with us confidently in English about their situations.

Descriptions of what a school library actually is vary from place to place and from country to country. Definitions have become important. For this reason, the IFLA *School Library Guidelines*, 2nd edition was

---

[1] Lo, P. & Chiu, D. Enhanced and Changing Roles of School Librarians under the Digital Age. *New Library World* 2015, *116*(11/12) 696–710.

published in 2016 (IFLA School Libraries Standing Committee, Dianne Oberg and Barbara Schultz-Jones (Eds.)).[2]

Indeed, school librarians working at local schools in developing countries face many problems, such as low literacy rates in the general population, the need to purchase books in many languages from overseas vendors, and often very little funding or administrative support (Gregorio, 2014).[3] Also, in many countries, there are no national policies for school libraries or a School Library Law, which compels school governing bodies and principals to have a library in their schools (Lo, et al., 2014[4]; Paton-Ash and Wilmot, 2015[5]). For this reason, most primary schools do not have libraries at all, while selective secondary schools might have very limited library facilities—for example, a classroom full of empty bookshelves with just few student tables, bare bulletin boards, and teacher desk.... According to Dr. Helen Boelens, "Some of my colleagues from developing countries would even describe a box of very old books, under lock and key in the principal's office as the 'School Library.' Not to mention the fact that training programs of school librarians vary greatly from country to country—that is why the IFLA has published a new set of guidelines."

Furthermore, there is very little expectation of school libraries in many regions (developing countries in particular), based on the system of education currently being offered, and with extremely limited facilities, or national/regional programs for training any librarians at any level. Even given the city of Hong Kong's vibrant and cosmopolitan economy, many classroom teachers amongst are assigned to manage the school libraries on a part-time basis, with the physical library regarded merely as a place for recreational reading, to be carried out in solitude.

This book does not intend to be comprehensive in breadth. Despite its limitations, this book has covered a large number of school librarians practicing "happily" and "successfully" in different school situations and environments in Australia, Bosnia and Herzegovina, Canada, Croatia, Democratic

[2] Oberg, D.; Schultz-Jones, B. Eds. *IFLA School Library Guidelines*. 2nd Ed.; International Federation of Library Associations and Institutions: Netherlands, 2015.

[3] Gregorio 2014 Quality School Libraries for Every Child: An International Concern. *OLA Q.* **2014,** *7*(1) 9+.

[4] Lo, et al. Attitudes and Self-Perceptions of School Librarians in Relations to Their Professional Practices: A Comparative Study between Hong Kong, Shanghai, South Korea, Taipei, and Japan. *Sch. Librar. Worldwide* 2014, *20*(1)51–69.

[5] Paton-Ash, M; Wilmot, D. Issues and Challenges Facing School Libraries in Selected Primary Schools in Gauteng Province, South Africa. *S. Afr. J. Edu.* 2015, *35*(1)1–10.

Republic of Congo, Hong Kong, Iceland, Ireland, Japan, Nepal, Nether-lands, Nigeria, Philippines, Serbia, Sweden, Thailand, Turkey, United States, Venezuela, and Zimbabwe. In examining the ways in which these school librarians live and work, in drastically diverse environments and conditions, and the policies, which result from them, this book provides a readable and unique account of how different policies, which regional educational poli-cies shape the operations and expectations of school libraries. It also offers a unique look at the many different faces of the profession of school librarian, in different cultural and language contexts.

Each of the school librarians mentioned in this book has a distinc-tively different educational and professional background, despite using the same job title; the scope of practice and their specialties vary greatly from each other. Nevertheless, this book has revealed a number of common trends and developments at both local and international levels that help address the issues and challenges faced by many school libraries practicing throughout the world. During their interviews, many school librarians have described how the roles of the school library and the school librarian have changed over time. The physical school library has been transformed from a place where children had access to books for reading and information, to a place in which the learner is the main focus. School libraries are no longer "depositories of information but transformational spaces" (Erikson and Markuson, 2007, p. ix)[6], where information is not only accessed, shared and stored, but challenged and created (Erikson and Markuson, 2007).[7]

The transformation into the schools' "learning hub" or "heart of learning" or "MakerSpaces" has resulted in the re-organization of library spaces, furniture, and collections. The school library also needs to be the center of the school, a dynamic and inviting place which has flex-ible, multifunctional spaces in which small and big groups can work and learn collaboratively together, more than one class can be accom-modated at the same time, digital media can be used and created, formal teaching can occur, and where the librarian can work collaboratively with teachers (Bolan, 2009[8]; Erikson and Markuson, 2007).

---

[6] Erikson, R.; Markuson, C. *Designing a School Library Media Centre for the Future.* 2nd Ed.; American Library Association: Chicago, Ill., 2007.

[7] Ibid.

[8] Bolan, K. *Teen Spaces: The Step-by-Step Library Makeover.* 2nd Ed.; American Library Association: Chicago, Ill., 2009.

Supportive to what Lo and Chiu (2015) pointed out in their earlier study, "Unlike other subject teachers, there is no "checklist-like" region-wide syllabus or school library standards that the school librarians can just observe and follow. This provides both opportunities and challenges, but these three school librarians have taken the better side of opportunities in the freedom to exercise their professional knowledge, skills, and judgments by exploring a variety of independent or collaborative teaching approaches, leading to a wide variation in library-oriented programs" (p. 707).

On the other hand, as pointed out by one of the school librarians who were interviewed for this book, "Not having to follow a region-wide syllabus or school library standards could also serve as a double-edged sword." Good professional practices and competencies come with experiences. Experience comes with age, persistent practice, continuous learning, as well as life lessons. Successful partnerships between school librarian and other subject teachers also take many years to develop. It also takes a great deal of effort to maintain this partnership under an ongoing basis. At their current workplace, all three librarians are given a great deal of freedom to exercise their professional knowledge, expertise as well as judgment to the fullest. On the other hand, inexperienced or entry-level librarians could find such "freedom" and "flexibility" difficult to cope and could easily lead to confusion and frustration. In summary:

1. The school librarian is often a "solo" position, and he/she is expected to work independently most of the time. For this reason, being able to understand the school culture and its management climate is considered highly important. Understanding the school culture also means being aware of the different resources and available options. Such an understanding would enable one to work within or around the school culture—so that library programs developed will be in line with the school's philosophy and practice. In order to carry out his/her work effectively, the strong support of the school leadership is essential.

2. To be able to effectively develop various strategies to extend the classroom curriculum and recreational reading through close educational partnerships with other subject teachers, school librarians require the necessary self-confidence and promotional skills similar to that of a marketing manager—that is, being

self-motivated, not afraid to take up extra work, innovative, and able to constantly think outside the box.

3. Teamwork is essential. Different subject teachers may have different working styles and needs; hence, one needs to be able to work collaboratively with other teachers across varying disciplines. For this matter, it is crucial that school librarians remain flexible, adaptive, creative, innovative and up-to-date (Lo and Chiu, 2015, p. 707).

Another major challenge faced by many school librarians is that if successful and fruitful collaborations to be achieved, a great deal of time and effort are needed in order to change the negative image built up by their predecessors. Nevertheless, their experiences and stories have demonstrated that sustainable successes can be achieved even with limited resources because possibilities and creativity could be limitless....

In addition to the practicing school librarians, the results of this book are of practical interest to education-policy makers, school administrators, as well as to educators with a vested interest in capitalizing the school library's potentials to positively affect students' academic achievements. Tertiary-level students who are majoring in Education (school librarianship in particular), at both undergraduate and postgraduate levels will find that this book enabling them to understand the reasoning behind the changes they are expected to face and implement. For example, recent advancement in information and communication technologies (ICTs) and the shift toward inquiry-based learning and its impacts on school libraries. The authors anticipate that the results of our study will also benefit them, preparing them to confront an uncertain educational world, whilst still retaining their enthusiasm for school librarianship.

Undoubtedly, the findings described in this book could further provide a strong basis for developing a pragmatic hypothesis and designing further quantitative studies in the next phase of further research by us and other researchers. Finally, it is hoped that this book could serve as an inspiration to those individuals who are considering a career in school librarianship.

To conclude this book, I would like to quote a statement made by a school librarian who took part in this interview book project—that summarizes the core value and contribution of this whole book:

"Bosnia and Herzegovina is a small country, and therefore, for a small school with a small library and a big heart like us, to be able to get out of the borders of our country and other cultural boundaries and being able to learn from school librarians of other countries would undoubtedly be of great worth. Your book featuring a series of interviews with school librarians practicing in many different parts of the world has provided with us the possibility that our work as school librarians and the work of our students gets noticed—both as bright examples in our own country and also around the world. Our interviews being included in your book is important because it gives a boost to our confidence and thereby enhancing our devotion and commitment to our work as school librarians. In short, your book has given us the opportunity for our school, and its library to open its doors to the world and present ourselves and our profession in the best possible way."—Ajdin Begic, School Librarian, Camil Sijarić Elementary School, Sarajevo, Bosnia and Herzegovina

INDEX